Practically Speaking
A Dictionary of Quotations
on Engineering, Technology and Architecture

About the Compilers

Carl C Gaither was born on 3 June 1944 in San Antonio, Texas. He has conducted research work for the Texas Department of Corrections and for the Louisiana Department of Corrections. Additionally he has worked as an Operations Research Analyst for ten years. He received his undergraduate degree (Psychology) from the University of Hawaii and has graduate degrees from McNeese State University (Psychology), North East Louisiana University (Criminal Justice), and the University of Southwestern Louisiana (Mathematical Statistics).

Alma E Cavazos-Gaither was born on 6 January 1955 in San Juan, Texas. She has previously worked in quality control, material control, and as a bilingual data collector. She is a Petty Officer First Class in the United States Navy Reserve. She received her associate degree (Telecommunications) from Central Texas College. Currently she is working on her BS Spanish.

Together they selected and arranged quotations for the books *Statistically Speaking: A Dictionary of Quotations* (Institute of Physics Publishing, 1996), *Physically Speaking: A Dictionary of Quotations on Physics and Astronomy* (Institute of Physics Publishing, 1997) and *Mathematically Speaking: A Dictionary of Quotations* (Institute of Physics Publishing, 1998).

About the Illustrator

Andrew Slocombe was born in Bristol in 1955. He spent four years of his life at Art College where he attained his Honours Degree (Graphic Design). Since then he has tried to see the funny side to everything and considers that seeing the funny side to engineering, technology and architecture has tested him to the full! He would like to thank Carl and Alma for the challenge!

The photograph of the authors on the previous page was taken by Kerry Yancey.

Practically Speaking
A Dictionary of Quotations on Engineering, Technology and Architecture

Selected and Arranged by

Carl C Gaither
and
Alma E Cavazos-Gaither

Illustrated by Andrew Slocombe

Institute of Physics Publishing
Bristol and Philadelphia

IOP Publishing Ltd has attempted to trace the copyright holders of all the quotations reproduced in this publication and apologizes to copyright holders if permission to publish in this form has not been obtained.

British Library Cataloguing-in-Publication Data
A catalogue record for this book is available from the British Library.

ISBN 0 7503 0594 0

Library of Congress Cataloging-in-Publication Data are available

Published by Institute of Physics Publishing, wholly owned by The Institute of Physics, London

Institute of Physics Publishing, Dirac House, Temple Back, Bristol BS1 6BE, UK
US Office: Institute of Physics Publishing, Suite 1035, The Public Ledger Building, 150 South Independence Mall West, Philadelphia, PA 19106, USA

Typeset in TEX using the IOP Bookmaker Macros
Printed in Great Britain by J W Arrowsmith Ltd, Bristol

I dedicate this book to my father
Clifford C. Gaither, LT. COL., USAF (RET)
Director of Physical Plant,
McNeese State University (1971–1991)

Carl C. Gaither

I dedicate this book to my son
Marcus Anthony Sutton, A.A.
Electronic Engineering Technology

Alma E. Cavazos-Gaither

If you can swing an axe, or wield a brush-hook,
  Or drive a stake, or drag a chain all day,
If you can scribble "figgers" in a note-book,
  Or shoot a range-pole half a mile away,
If you can sight a transit or a level,
  Or move a target up and down a rod,
If you fear neither man nor devil,
  And know yourself and trust the living God;

If you can wade a swamp, or swim a river,
  Nor fear the deeps, nor yet the dizzy heights,
If you can stand the cold without a shiver,
  And take the Higgins' ink to bed o' nights,
If you can turn a thumb-screw with your fingers,
  When every digit's like a frozen thumb,
If you can work as long as daylight lingers
  And not complain, nor think you're going some;

If you can sight through tropic-heat's refraction,
  Or toil all day beneath a blistering sun,
If you can find a sort of satisfaction
  In knowing that you've got a job well done,
If you can be an Esquimau and nigger
  And try to be a gentleman, to boot,
If you can use a "guessing stick" to figger,
  And know a coefficient from a root;

If your calculus and descriptive are forgotten,
  And your algebra just serves you fairly well,
If your drafting and your lettering are rotten,
  And your Trautwine's always handy by to tell,
If you can close a traverse without fudgin',
  Or check a line of levels by a foot,
If you can set a slope-steak, just by judgin',
  And never kick a tripod with your foot;

If you can run a line where you are told,
  And make it stay somewhere upon the map,
If you can read your notes when they get cold,
  And know that contours mustn't ever lap,
If you can line a truss or tap a river,
  Or make a surly foreman come across,
If you can take an order, well as give it,
  And not have secret pity for the boss;

If you can climb a stool and not feel lowly,
    Nor have your head turned by a swivel chair,
If you can reach your judgments slowly,
    And make your rulings always just and fair,
If you can give yourself and all that's in you,
    And make the others give their own best, too,
If you can handle men of brawn and sinew,
    And like the men and make 'em like you too;

If you can't boast a college education,
    Or, if you've got a sheep-skin, can forget,
If you get a living wage for compensation,
    And give a little more than what you get,
If you can meet with triumph and disaster,
    And treat them without favor, nor with fear,
You'll be a man—and your own master,
    But—what is more—you'll be an ENGINEER.

Randolph, Robert Isham
IF

In J.A.L. Waddell, Frank W. Skinner and Harold E. Wessman (Editors)
Vocational Guidance in Engineering Lines, First Edition (pp. 19 and 24)

CONTENTS

PREFACE

With so many well-prepared books of quotations on the market is another book of quotations necessary? We and our publisher agreed that there was a need since the standard dictionaries of quotations, for whatever cause, are conspicuous by their lack of entries devoted to quotations on engineering, technology and architecture. *Practically Speaking* fills that need.

The understanding of the history, the accomplishments and failures, and the meanings of engineering, technology and architecture requires a knowledge of what has been said by the authoritative and the not so authoritative engineers, architects, technicians, philosophers, novelists, playwrights, poets, scientists and laymen. Because of the multidisciplinary interrelationships that exist between engineering, technology and architecture it is virtually impossible for individuals to keep abreast of the literature outside of their own particular specialization. With this in mind, *Practically Speaking* assumes a particularly important role as a guide to what has been said in the past through to the present about these disciplines.

Practically Speaking was designed as an aid for the general reader who has an interest in these topics as well as for the experienced engineer, architect or technician. The general reader with no knowledge of engineering, technology or architecture who reads *Practically Speaking* can form a pretty accurate picture of these fields. Students can use the book to increase their understanding of the complexity and richness that exists within and between these disciplines. Finally, the experienced professional will find *Practically Speaking* useful as a source of quotes for use in the classroom, in papers and in presentations. We have striven to compile the book so that any reader can easily and quickly access the wit and wisdom that exists, and a quick glance through the table of contents will show the variety of topics discussed.

A book of quotations, even as restricted in scope as *Practically Speaking* is, can never be complete. Many quotations worthy of entry have, no doubt, been omitted because we did not know of them. However, we have tried to make it fairly comprehensive and have searched far and wide for the material. If you are aware of any quotes that should be included please send them in for the second edition.

Quite a few of the quotations have been used frequently and will be recognized while others have probably not been used before. All of the quotations included in *Practically Speaking* were compiled with the hope that they will be found useful. The authority for each quotation has been given with the fullest possible information that we could find so as to help you pinpoint the quotation in its appropriate context or discover more quotations in the original source. When the original source could not be located we indicated where the quote was found. Sometimes, however, we only had the quote and not the source. When this happened we listed the source as unknown and included the quotation anyway so that it would not become lost with time.

It has been our good fortune to benefit from the comments and suggestions made by the reviewers who are authorities in their field. We wish to thank them for their very important contributions and their suggestions have been listened to with particular attention. However, the reviewers frequently contradicted each other. Some said we had omitted significant topics and authors, others said we had too many topics and authors, and still others said the topics and authors we had were fine. Thus, the final form of the book represents a blend of their suggestions.

How to Use This Book
1. A quotation for a given subject may be found by looking for that subject in the alphabetical arrangement of the book itself. This arrangement will be approved, we believe, by the reader in making it easier to locate a quotation. To illustrate, if a quotation on "ideas" is wanted, you will find twenty-four quotations listed under the heading Idea. The arrangement of quotations in this book under each subject heading constitutes a collective composition that incorporates the sayings of a range of people.
2. To find all the quotations pertaining to a subject and the individuals quoted use the SUBJECT BY AUTHOR INDEX. This index will help guide you to the specific statement that is sought. A brief extract of each quotation is included in this index.
3. It will be admitted that at times there are obvious conveniences in an index under author's names. If you recall the name appearing in the attribution or if you wish to read all of an individual author's contributions that are included in this book then you will want to use the AUTHOR BY SUBJECT INDEX. Here the authors are listed alphabetically along with their quotations. The birth and death dates are provided for the authors whenever we could determine them.

Thanks

It is never superfluous to say thanks where thanks are due. Firstly, we want to thank Jim Revill and Al Troyano, of Institute of Physics Publishing, who have assisted us so very much with our books. Next, we thank the following libraries for allowing us to use their resources: the Jesse H. Jones Library and the Moody Memorial Library, Baylor University; the main library of the University of Mary-Hardin Baylor; the main library of the Central Texas College; the Undergraduate Library, the Engineering Library, the Law Library, the Physics-Math-Astronomy Library, and the Humanities Research Center, all of the University of Texas at Austin. Again, we wish to thank Joe Gonzalez, Matt Pomeroy, Chris Braun, Ken McFarland, Craig McDonald, Kathryn Kenefick, Brian Camp, Robert Clontz, and Gabriel Alvarado of the Perry-Castañeda Library for putting up with us when we were checking out the hundreds of books. Finally, we wish to thank our children Maritza, Maurice, and Marilynn for their assistance in finding the books we needed when we were at the libraries.

A great amount of work goes into the preparation of any book. When the book is finished there is then time for the editors and authors to enjoy what they have written. It is hoped that this book will stimulate your imagination and interests in matters about engineering, technology and architecture, and this hope has been eloquently expressed by Helen Hill:

> If what we have within our book
>   Can to the reader pleasure lend,
> We have accomplished what we wished,
>   Our means have gained our end.

In Llewellyn Nathaniel Edwards
A Record of History and Evolution of Early American Bridges (p. xii)

Carl Gaither
Alma Cavazos-Gaither
September 1998

ANALYSIS

Allen, Roy George Douglas
Not even the most subtle and skillful analysis can overcome completely
the unreliability of basic data . . .

Statistics for Economists
Chapter I, section 1.4 (p. 14)

Amiel, Henri Frédéric
Minds accustomed to analysis never allow objections more than half-
value, because they appreciate the variable and relative elements which
enter in.

Journal Intime (p. 166)

Analysis kills spontaneity. The grain once ground into flour springs and
germinates no more.

Journal Intime (p. 325)

Keeney, Ralph
Raiffa, Howard
. . . be wary of analysts that try to quantify the unquantifiable.

Decisions with Multiple Objectives: Preferences and Value Trade-Off (p. 12)

Poe, Edgar Allan
The analytical power should not be confounded with simple ingenuity;
for while the analyst is necessarily ingenious, the ingenious man is often
remarkably incapable of analysis.

The Complete Edgar Allan Poe Tales
The Murders in the Rue Morgue (p. 247)

1

Reed, E.G.
. . . the primary purpose of analysis is to simplify a problem rather than
solve it . . .

Machine Design
Developing Creative Talent (p. 144)
November 1954

Whitehead, Alfred North
It requires a very unusual mind to undertake the analysis of the obvious.

Science and the Modern World
The Origins of Modern Science (p. 4)

Logic hasn't wholly dispelled the society of witches and
prophets and sorcerers and soothsayers.
Raymond F. Jones – (See p. 178)

ANSWER

Adams, Henry Brooks
Unintelligible answers to insoluble problems.

In Bert Leston Taylor
The So-Called Human Race (p. 154)

Arnold, John E.
. . . there is no one right answer to creative problems.

In Sidney J. Parnes and Harold F. Harding
A Source Book for Creative Thinking
Useful Creative Techniques (p. 252)

Baez, Joan
. . . hypothetical questions get hypothetical answers.

Daybreak
What Would You Do If (p. 134)

Berkeley, Edmund C.
The moment you have worked out an answer, start checking it—it probably isn't right.

Computers and Automation
Right Answers—A Short Guide for Obtaining Them (p. 20)
September 1969

Dobie, J. Frank
Putting on the spectacles of science in expectation of finding the answer to everything looked at signifies inner blindness.

The Voice of the Coyote
Introduction (p. xvi)

Hodnett, Edward
You have to ask a precise question to get a precise answer.

The Art of Problem Solving (p. 37)

Woodson, Thomas T.
. . . it is safe money to wager that an unproven answer is wrong.

Introduction to Engineering Design (p. 240)

ARCH

Fergusson, James
. . . an arch never sleeps . . .

History of Indian and Eastern Architecture
Volume 1, Chapter II (p. 210)

ARCHITECT

Alexander, Daniel Asher

'You are a builder, I believe?'

'No, Sir; I am not a builder; I am an architect.'

'Ah well, builder or architect, architect or builder—they are pretty much the same I suppose?'

'I beg your pardon; they are both totally different.'

'Oh, indeed! Perhaps you will state wherein this difference consists.'

'An architect, sir, conceives the design, prepares the plan, draws out the specification—in short, supplies the mind. The builder is merely the machine; the architect the power that puts the machine together and sets it going.'

'Oh, very well, Mr. Architect, that will do. A very ingenious distinction without a difference. Do you happen to know who was the architect of the Tower of Babel?'

'There was no architect, sir. Hence the confusion.'

Sir James Courlett (Interrogator)
In H.M. Colvin
A Biographical Dictionary of English Architects

Carr-Saunders, A.M.
Wilson, P.A.

The architect therefore is not only a professional man but also an artist, and he shares in some measure both the problems and the qualities of temperament commonly associated with artists. Thus he lives in a world which is dominated by fashion and split up into cliques and coteries, and he brings to the consideration of professional business something of the outlook which characterizes his interest in the subject-matter of his art.

The Professions (p. 184)

Cowper, William
But 'tis not timber, lead and stone,
An architect requires alone,
To finish a fine building—
The palace were but half complete,
If he could possibly forget
The carving and the gilding.

The Complete Poetical Works of William Cowper
Friendship, l. 163–168

de Bergerac, Cyrano
The Architector that built my Prison, having made my Entries into it, did not bethink himself of making one Outlet.

The Comical History of the States and Empires of the Worlds of the Moon and Sun
The History of the World of the Sun (pp. 25–6)

Esar, Evan
The architect makes an old house look better just by talking about the cost of a new one.

20,000 Quips and Quotes

The architect who always talks shop is probably suffering from an edifice complex.

20,000 Quips and Quotes

Keate, G.
When *Troy* was built, you recollect
I dabbled as an *Architect*;
A very sorry one, you'll say,
But worse since then have come in play,
And of the art I've understood
Enough, to do more harm than good,

The Distressed Poet
Canto the Second, l. 332–337

Le Corbusier [Jeanneret-Gris, Charles Édouard]
The Architect, by his arrangement of forms, realizes an order which is a pure creation of his spirit . . . it is then that we experience the sense of beauty.

Towards a New Architecture (p. 7)

Longfellow, Henry Wadsworth
. . . The architect
Built his great heart into these sculptured stones,
And with him toiled his children, and their lives

Were builded, with his own, into the walls,
As offerings unto God.

The Complete Poetical Works of Longfellow
Christus: A Mystery
The Golden Legend
In the Cathedral (p. 431)

Mangan, J.C.
"Architect! my handsome country villa
Yesterday took fire, and nought could save it.
It now lies a ruin!"—"Allah-el-illah!
Fire, like Air, will find or force expansion—
Fire must burn, and woodwork may not brave it!
But—I'll build thee a far handsomer mansion."
—"Thanks, good Architect! The cost may make me
Poorer, but, Inshallah! 'twill not break me."

Poems of James Clarence Mangan
The Worst Loss
Stanza II, l. 9–16

Milton, John
The hasty multitude
Admiring enter'd, and the work some praise,
And some the architect: his hand was known
In heaven by many a tower'd structure high,
Where scepter'd angels held their residence,
And sat as princes.

Paradise Lost
Book I, l. 730–735

North, Roger
For a profest architect is proud, opiniative and troublesome, seldome at hand, and a head workman pretending to the designing part, is full of paultry vulgar contrivances; therefore be your owne architect, or sitt still.

In H.M. Colvin
A Biographical Dictionary of English Architects, 1600–1840 (p. 13)

Nuttgens, Patrick
The challenge for the modern architect is the same as the challenge for all of us in our lives: to make out of the ordinary something out-of-the-ordinary.

BBC TV program
Architecture for Everyman
Reprinted in
The Listener
1 March 1979

Ruskin, John
An architect should live as little in cities as a painter. Send him to our hills, and let him study there what nature understands by a buttress, and what by a dome.

True and the Beautiful
Part 3
The Lamp of Power (p. 129)

Shipman, T.
An *Architect* should chiefly try
To please the *Owner's* Mind and Eye . . .

Carolina: or Loyal Poems (p. 234)
To the Reader of the following Poem

Shute, John
. . . an Architecte must be sharpe of understandinge and both quicke and apte to conceiue the trewe Instructions and meaninges of them that have written thereof: and must also be a perfect distributor of the great misteries that he hath perceued and experymented, that playnlye and briefly he may discusse and open demonstrations of that which shallbe done . . .

The First and Chief Groundes of Architecture
folio iii

It belongeth also to an Architect, to haue sight in Philosophie, which teaching to be of a noble courage as Virtuuius saith, and also gentil, curtious, faithfull and modest, not geuen to auarice and filthy lucre, as not to be troubled or corrupted with rewardes or giftes, but with grauity and Sagenes to coceiue al honor and dignity in al things conseruinge his good name and estimation. Let him also take a charge of workes in hand, being desired and not desirous of workers.

The First and Chief Groundes of Architecture
folio iii

Soane, Sir John
The business of the architect is to make the designs and estimates, to direct the works, and to measure and value the different parts, he is the intermediary agent between the employer, whose honour and interests he is to study, and the mechanic whose rights he is to defend. His situation implies great trust; he is responsible for the mistakes, negligencies and ignorancies of those he employs, and, above all, he is to take care that the workman's bills do not exceed his own estimate. If these are the duties of an architect, with what propriety can his situation and that of the builder or contractor be united?

In Arthur T. Bolton
Life and Work a Century Ago: an outline of the career of Sir John Soane (p. 5)

Vitruvius
. . . architects who without culture aim at manual skill cannot gain a prestige corresponding to their labours, while those who trust to theory and literature obviously follow a shadow and not reality. But those who have mastered both, like men equipped in full armour, soon acquire influence and attain their purpose.

Vitruvius on Architecture
Volume I
Book I, Chapter I, 2 (p. 7)

Wightwick, George
It is presumed your primary object in securing the services of an Architect involves the recognition of his pretensions as an Artist. The ordinary *Builder* may construct the edifice required: you apply to an *Architect* for the superadded graces of correct design and suitable decoration . . . *In matters of Taste* he engages to give you what he conceives to be correct, and to the amount only which your means allow, and *not* to sacrifice without reluctance his repute as an Artist to your individual wishes, not to suffer under your censure for limiting his decorations to their just proportion in the general outlay.

Letter to prospective clients, circa 1825
Royal Institute of British Architects Proceedings, New Series
Volume VII, 1891 (p. 161)

Wright, Frank Lloyd
The physician can bury his mistakes, but the architect can only advise his client to plant vines.

New York Times Magazine
Frank Lloyd Wright and His Art (p. 47)
4 October 1953

Any good architect is by nature a physicist as a matter of fact, but as a matter of reality, as things are, he must be a philosopher and a physician.

Frank Lloyd Wright: An Autobiography
The Character of Form (p. 380)

ARCHITECTURE

Bacon, Francis
Houses are built to live in, and not to look on: therefore let use be preferred before uniformity.

In Brian Vickers (Editor)
Francis Bacon
Essays
Of Buildings (p. 427)

He that builds a fair house upon an ill seat, committeth himself to prison.

In Brian Vickers (Editor)
Francis Bacon
Essays
Of Buildings (p. 427)

Buchanan, R.W.
Thus I taught them architecture,—
How to hew the rocks and fashion
Monuments that stand for ever
In despite of God and Time.

The Complete Poetical Works of Robert Buchanan
The Devil's Case
XVI, l. 1377–1380

Carr-Saunders, A.M.
Wilson, P.A.
Architecture differs from every other profession . . . in that the technique contains an aesthetic element. Indeed, the aesthetic element is fundamental; and no matter how complex the science of building construction is or may become, the architect is only concerned with that science in order to apply it to aesthetic purposes.

The Professions (p. 184)

Dimnet, Ernest
Architecture, of all the arts, is the one which acts the most slowly, but the most surely, on the soul.

What We Live By
Chapter VII (p. 141)

Emerson, Ralph Waldo
The Gothic cathedral is a blossoming in stone subdued by the insatiable demand of harmony in man. The mountain of granite blooms into an eternal flower.

Essays
History (p. 12)

Goethe, Johann Wolfgang von
I have found a paper of mine . . . in which I call architecture 'petrified music'.

In Johann Peter Eckermann
Conversations with Goethe
23 March 1829

Johnson, Philip
Architecture is the art of how to waste space.

New York Times
Ideas and Men
Section E, page 9
27 December 1964

Le Corbusier [Jeanneret-Gris, Charles Édouard]
There is one profession and one only, namely architecture, in which progress is not considered necessary, where laziness is enthroned, and in which the reference is always to yesterday.

Towards A New Architecture (p. 101)

Lytton, E.R.B.
Thither once came a traveller who had read
Marcus Vitruvius Pollio, and had all
The terms of architecture in his head,
Apophyge, and plinth, and astragal.

The Titlark's Nest
A Parable, l. 25–28

Nietzsche, Friedrich
. . . architecture is a kind of oratory in forms, sometimes persuading or even flattering, sometimes simply commanding.

Twilight of the Idols
Raids of an Untimely Man
11 (p. 57)

Piper, Robert J.
Architecture is the art and science of designing buildings and the spaces
between them.

Opportunities in an Architecture Career (p. 13)

Ruskin, John
Ornamentation is the principal part of architecture, considered as a
subject of fine art.

True and the Beautiful
Part 4
Sculpture (p. 208)

Architecture is the work of nations . . .

True and the Beautiful
Part 4
Sculpture (p. 187)

The value of Architecture depended on two distinct characters:—the one,
the impression it receives from human power; the other, the image it
bears of the natural creation.

True and the Beautiful
Part 3
The Lamp of Beauty (p. 130)

We may live without her [architecture], and worship without her, but
we cannot remember without her.

True and the Beautiful
Part 3
The Lamp of Memory (p. 140)

Schelling, Friedrich
Architecture is frozen music.

Philosophie der Kunst (p. 576)

Smart, C.
Ma'm, architecture you're not skill'd in,
I don't approve your way of building;
In this there's nothing like design,
Pray learn the use of Gunter's line.

The Poetical Works of Christopher Smart
The Blockhead and Beehive
Fable X, l. 59–62

Wright, Frank Lloyd
The only thing wrong with architecture are the architects.

In Evan Esar
20,000 Quips and Quotes

Wyatt, Mrs. James
You must be aware that Architecture is the profession of a Gentleman, and that none is more lucrative when it is properly attended to.

Letter to son Philip, 1808
Egerton Manuscript
3515

You have to ask a precise question to get a precise answer.
Edward Hodnett – (See p. 3)

ASSUMPTION

Bell, Eric T.
Pick the assumptions to pieces till the stuff they are made of is exposed to plain view—this is the cardinal rule for understanding the basis of our beliefs.

The Search for Truth (p. 25)

Donghia, Angelo
Assumption is the mother of screw-up . . .

New York Times
Behind Angelo Donghia's Gray Flannel Success
Section C, page 6
20 January 1983

Siegel, Eli
An assumption, as such, is really not more daring than the facts.

Damned Welcome (p. 21)

AUTHORITY

da Vinci, Leonardo
Anyone who in discussion relies upon authority uses, not his understanding, but rather his memory.

The Literary Works of Leonardo da Vinci
Volume II, 1159 (p. 241)

BEAUTY

Bacon, Francis
There is no excellent beauty that hath not some strangeness in the proportion.

In Brian Vickers (Editor)
Francis Bacon
Essays
Of Beauty (p. 425)

Bridges, Robert Seymour
For beauty being the best of all we know
Sums up the unsearchable and secret aims
Of nature.

The Growth of Love
VIII

Dickinson, Emily
Beauty is not caused, It is.

Further Poems
Number 57

Emerson, Ralph Waldo
Beauty rests on necessities. The line of beauty is the line of perfect economy.

The Works of Ralph Waldo Emerson
Volume VI
Conduct of Life
Beauty

We ascribe beauty to that which is simple; which has no superfluous parts; which exactly answers its end.

The Works of Ralph Waldo Emerson
Volume VI
Conduct of Life
Beauty

Holmes, Oliver Wendell
Beauty is the index of a larger fact than wisdom.

The Professor at the Breakfast Table
Chapter 2 (p. 31)

Hugo, Victor
The beautiful is as useful as the useful. More so, perhaps.

Les Misérables
Fantaine
Book I, Chapter 6

Keats, John
A thing of beauty is a joy forever: . . .

The Poetical Works of John Keats
Endymion
Book I, l. 1

Beauty is truth, truth beauty . . .

The Poetical Works of John Keats
Ode on a Grecian Urn

Penrose, Roger
A beautiful idea has a much greater chance of being a correct idea than an ugly one.

The Emperor's New Mind
Chapter 10 (p. 421)

Sartre, Jean-Paul
The real is never beautiful. Beauty is a value which applies only to the imaginary and which entails the negation of the world in its essential structure.

In T. Dobzhansky
The Biology of Ultimate Concern
Chapter 5 (p. 102)

Wallace, Lew
. . . beauty is altogether in the eye of the beholder . . .

The Prince of India
Book III, Chapter vi (p. 122)

Wilde, Oscar
Beauty is a form of Genius—is higher, indeed, than Genius, as it needs no explanation.

The Picture of Dorian Gray
Chapter 2

BRIDGE

Andric, Ivo
When the angels saw how unfortunate men could not pass those abysses and ravines to finish the work they had to do, but tormented themselves and looked in vain and shouted from one side to the other, they spread their wings above those places and men were able to cross. So men learned from the angels of God how to build bridges, and therefore, after fountains, the greatest blessing is to build a bridge.

The Bridge on the Drina
Chapter XVI (pp. 208–9)

Broun, Heywood
Men build bridges and throw railroads across deserts, and yet they contend successfully that the job of sewing on a button is beyond them. Accordingly, they don't have to sew buttons.

Seeing Things at Night
Holding a Baby (p. 168)

Brown, Allen
Bridges, like humans, have birthdays. They are reckoned from the day they are opened to traffic. But the conception of a bridge is unnatural; bridging a waterway is an act against Nature herself—an act that Nature frequently contests and sometimes does not permit.

Golden Gate (p. 3)

Magna Carta
No township or subject shall be compelled to make bridges at river banks, except those who by ancient usage are legally bound to do so.

In J.C. Dickinson
The Great Charter
Chapter 23 (p. 22)

18

McGonagall, William
Oh! ill fated Bridge of the Silv'ry Tay,
I must now conclude my lay
By telling the world fearlessly and without the least dismay,
That your central girders would not have given way,
At least many sensible men do say,
Had they been supported on each side with buttresses,
At least many sensible men confesses,
For the stronger we our houses do build,
The less chance of being killed.

Poetic Gems
The Tay Bridge Disaster (p. 92)

Petroski, Henry
Designing a bridge or any other large structure is not unlike planning a trip or vacation. The end may be clear and simple: to go from here to there. But the means may be limited only by our imagination.

To Engineer is Human (p. 64)

Roebling, John
The contemplated work, when constructed in accordance with my designs, will not only be the greatest Bridge in existence, but it will be the greatest engineering work of this continent, and of the age. Its most conspicuous features, the great towers, will serve as landmarks to the adjoining cities, and they will be entitled to be ranked as national monuments.

Report to the New York Bridge Company
1867

Schuyler, Montgomery
It so happens that the work which is likely to be our most durable monument, and to convey some knowledge of us to the most remote posterity, is a work of bare utility; not a shrine, not a fortress, not a palace but a bridge.

Harper's Weekly
The Bridge as a Monument (p. 326)
Volume XXVII, Number 1379, 27 May 1883

Steinman, D.B.
Between two towers soaring high
A parabolic arc is swung
To form a cradle for the stars;
And from this curve against the sky
A span of gleaming steel is hung—
A highway of speeding cars.

Between the cable and the span
  A web of silver strands is spaced,
  With sky above and ships below
In human dream was born the plan
  Of strength and beauty interplaced—
  A harp against the sunset glow!

American Engineer
SUSPENSION BRIDGE (p. 33)
22–28 February 1953

Woodson, Thomas T.
Poor arithmetic will make the bridge fall down just as surely as poor physics, poor metallurgy, or poor logic will.

Introduction to Engineering Design (p. 245)

BUILD

Longfellow, Henry Wadsworth
Michael A: Ah, to build, to build!
That is the noblest art of all the arts.
Painting and sculpture are but images,
Are merely shadows cast by outward things
On stone or canvas, having in themselves
No separate existence. Architecture,
Existing in itself, and not seeming
A something it is not, surpasses them
As substance shadow.

The Complete Poetical Works of Longfellow
Michael Angelo
III, San Silvestro

Ruskin, John
Therefore when we build, let us think that we build [public edifices]
forever. Let it not be for present delight, nor for present use alone, let
it be such work as our descendants will thank us for, and let us think,
as we lay stone to stone, that a time is to come when those stones will
be held sacred because our hands have touched them, and that men will
say as they look upon the labor and wrought substance of them, "See!
this our fathers did for us."

True and the Beautiful
Part 3
The Lamp of Memory (pp. 142–3)

Shakespeare, William
When we mean to build,
We first survey the plot, then draw the model;
And when we see the figure of the house,
Then must we rate the cost of the erection.

Henry IV
Part II, Act I, scene 3, l. 41–44

Unknown

Those who personally dominate are heroes for the hour; those who build are immortal.

The Journal of Engineering Education
Volume 30, Number 3, November 1939 (p. 314)

Wooten, Henry

In *Architecture* as in all other *Operative* Arts, the *end* must direct the *Operation*.

The *end* is to build well.

Well building hath three Conditions.

Commodities, Firmenes, and *Delight*.

·*The Elements of Architecture*
The I. part (p. 1)

BUILDER

Egerton-Warburton, R.E.
The plan reduced from small to less to make his house compactor,
The Builder, his own Architect, became his own Contractor.

Caution And Economy, l. 1–2

Hammurabi
If a builder of a house for any one and do not build it solid; and the house, which he has built, fall down and kill the owner; one shall put that builder to death.

The Codes of Hammurabi and Moses
229 (p. 93)

Longfellow, Henry Wadsworth
In the elder days of Art,
  Builders wrought with greatest care
Each minute and unseen part;
  For the Gods see everywhere.

The Complete Poetical Works of Longfellow
The Builders

Ruskin, John
No person who is not a great sculptor or painter *can* be an architect. If he is not a sculptor or painter, he can only be a *builder*.

True and the Beautiful
Part 4
Sculpture (p. 209)

Seneca
Believe me, that was a happy age, before the days of architects, before the days of builders!

The Epistles of Seneca
Epis. xc, sec. 9

Tredgold, Thomas
. . . the strength of a building is inversely proportional to the science of the builder.

Practical Essay on the Strength of Cast Iron and Other Metals

BUILDING

Gloag, John
Architecture cannot lie, and buildings, although inanimate, are to that extent morally superior to men.

Presentation
Royal Society of Arts
20 March 1963
The Significance of Historical Research in Architectural and Industrial Design

Ruskin, John
. . . we require from buildings, as from men, two kinds of goodness: first, the doing their practical duty well: then that they be graceful and pleasing in doing it; which last is itself another form of duty.

The Stones of Venice
Volume I, Chapter II, section 1 (p. 39)

Better the rudest work that tells a story or records a fact, than the richest without meaning. There should not be a single ornament put upon great civic buildings, without some intellectual intention.

True and the Beautiful
Part 3
The Lamp of Memory (p. 142)

CALCULATION

Birns, Harold
One might risk establishing the following mathematical formula for bribery . . . namely OG = PLR × AEB: The opportunity for graft equals the plethora of legal requirements multiplied by the number of architects, engineers and builders.

New York Times
City Acts to Unify Inspection Rules (p. 43)
Column 8, 2 October 1963

Boswell, James
It is wonderful when a calculation is made, how little the mind is actually employed in the discharge of any profession.

The Life of Samuel Johnson
6 April 1775

de Saint-Exupéry, Antoine
The sailing vessel itself was once a machine born of the calculations of engineers, yet it does not disturb our philosophers. The sloop took its place in the speech of men. There is a poetry of sailing as old as the world. There have always been seamen in recorded time. The man who assumes that there is an essential difference between the sloop and the airplane lacks historic perspective.

Wind, Sand and Stars
The Tool (p. 72)

Weinberg, Gerald M.
BEFORE YOU CAN COUNT ANYTHING, YOU'VE GOT TO KNOW SOMETHING

Rethinking Systems Analysis and Design (p. 32)

Wittgenstein, Ludwig
The process of *calculating* brings about just this intuition. Calculation is not an experiment.

Tractatus Logico Philosophicus
6.2331 (p. 171)

CAUSE AND EFFECT

Akenside, Mark
Give me to learn each secret cause;
Let number's figure's, motion's laws
Reveal'd before me stand;
These to great Nature's scenes apply,
And round the Globe, and thro' the sky,
Disclose her working hand.
The Poetical Works of Mark Akenside and John Dyer (l. 19–24)
Hymn to Science in Works of the English Poets (p. 357)

Arthur, T.S.
Only a few look at causes, and trace them to their effects.
Ten Nights in a Bar Room and What I Saw There
Night the Fifth

Atherton, Gertrude
The law of cause and effect does not hide in the realm of the unexpected when intelligent beings go looking for it.
Senator North
Book II, XXI

da Vinci, Leonardo
There is no result in nature without a cause; understand the cause and you will have no need for the experiment.
Leonardo da Vinci's Notebooks
Of the Intellectual Life (p. 54)

Dryden, John
Happy the man, who studying Nature's laws,
Through known effects can trace the secret cause—

His mind, possessing in a quiet state,
Fearless of fortune and resigned to fate.

The Poetical Works of Dryden
Translation of Virgil
The Second Book of the Georgics, l. 701–704

Froude, James Anthony
Every effect has its cause.

Short Studies on Great Subjects
Calvinism (p. 12)

Holmes, Oliver Wendell
But he who, blind to universal laws,
Sees but effects, unconscious of the causes,—

The Complete Poetical Works of Oliver Wendell Holmes
A Metrical Essay

Jackson, Hughlings
The study of the causes of things must be preceded by the study of things caused.

In W.I.B. Beveridge
The Art of Scientific Investigation
Preparation (p. 10)

CHAOS

Adams, Henry Brooks
In plain words, Chaos was the law of nature; Order was the dream of man.

The Education of Henry Adams
The Grammar of Science (p. 451)

Dylan, Bob
Chaos is a friend of mine.

Newsweek
The Two Lives of Bob Dylan (p. 93)
9 December 1985

Kettering, Charles Franklin
Developmental work is always a slightly organized chaos.

In T.A. Boyd
Professional Amateur (p. 71)

Stewart, Ian
. . . chaos is 'lawless behavior governed entirely by law'.

Does God Play Dice? (p. 17)

COMMON SENSE

Compton, Karl Taylor
We cannot get far by trying to impose an engineering education, however excellent it may be, on a young man of mediocre ability or one temperamentally unfitted for technical or administrative work. The idea reminds me of an experience which my sister had . . . in India. She had engaged a native electrician to install some new fixtures in her house, but he seemed particularly stupid and kept coming to her for instructions. Finally, in exasperation she said to him: "Why do you come asking questions all the time? Why don't you use your common sense?" "Madam," he replied gravely, "common sense is a rare gift of God. I have only a technical education."

A Scientist Speaks (p. 51)

Cross, Hardy
Common sense is only the application of theories which have grown and been formulated unconsciously as a result of experience.

Engineers and Ivory Towers
For Man's Use of God's Gifts (p. 107)

The idea that common sense is a gift of the gods is overdone. Some men will never have common sense in engineering problems; but it can be developed to some extent by those who work hard and hopefully can then repeatedly look over the whole field in which they have worked. They must try to see the hills and the valleys, try to appreciate what parts are important and what parts are less important, try to be synthetic as well as analytic, to give due attention to probability, to develop some sense of relative importance. To these men will come in time what seems an intuitive faculty of getting answers. Common sense provides a rapid qualitative approach to problems.

Engineers and Ivory Towers
For Man's Use of God's Gifts (p. 108)

Roosevelt, Franklin D.
It is common sense to take a method and try it. If it fails, admit it frankly and try another. But above all, try something.

Oglethorpe University, speech, 22 May 1932

Russell, Bertrand
Common sense, however it tries, cannot avoid being surprised from time to time. The aim of science is to save it from such surprises.

In Jean-Pierre Luminet
Black Holes (p. 182)

Whitehead, Alfred North
Now in creative thought common sense is a bad master. Its sole criterion for judgment is that the new ideas shall look like the old ones. In other words it can only act by suppressing originality.

An Introduction to Mathematics
Chapter 11 (p. 116)

COMMUNICATION

Casimir, Hendrik B.G.
There exists today a universal language that is spoken and understood almost everywhere: it is Broken English. I am not referring to Pidgin-English—a highly formalized and restricted branch of B.E.—but to the much more general language that is used by the waiters in Hawaii, prostitutes in Paris and ambassadors in Washington, by businessmen from Buenos Aires, by scientists at international meetings and by dirty-postcard peddlers in Greece—in short, by honorable people like myself all over the world . . . The number of speakers of Broken English is so overwhelming and there are so many for whom B.E. is almost the only way of expressing themselves—at least in certain spheres of activity—that it is about time that Broken English be regarded as a language in its own right.

Haphazard Reality
Chapter 4 (p. 122)

Chargaff, Erwin
There is no real popularization possible, only vulgarization that in most instances distorts the discoveries beyond recognition.

Perspectives in Biology and Medicine
Bitter Fruits from the Tree of Knowledge (p. 491)
Volume 16, Summer 1973

I should like to find a way of discouraging unnecessary publications, but I have not found a solution, save the radical one . . . that all scientific papers be published anonymously.

Perspectives in Biology and Medicine
In Praise of Smallness (p. 383)
Volume 23, Spring 1980

Dubos, René
. . . a scientific paper should never try to make more than one point.

In B.D. Davis
Perspectives in Biology and Medicine
Two Perspectives (p. 38)
Volume 35, Autumn 1991

Gastel, Barbara
Every master's thesis or doctoral dissertation should be accompanied by a lay summary or press release written by the graduate student (with the guidance, if possible, of a science writing instructor or public information officer at the student's institution).

Earth and Life Science Editing
Volume 24, 1985 (p. 3)

Gibbs, Willard
. . . science is, above all, communication.

Attributed in H.N. Parton
Science is Human (p. 11)

Moravcsik, M.J.
New theories, when first proposed, may appear on the first page of the *New York Times*, but their demise, a few years later, never makes even page 68.

Research Policy
Volume 17, 1988 (p. 293)

Seifriz, W.
Our scientific conferences are a hodgepodge of trivia. The conversation is that of men on the defensive.

Science
A New University (p. 89)
Volume 120, Number 3106, 9 July 1954

Ziman, J.M.
Although the best and most famous scientific discoveries seem to open whole new windows of the mind, a typical scientific paper has never pretended to be more than another piece in a large jig-saw— not significant in itself but as an element in a grander scheme. This technique, of soliciting many modest contributions to the vast store of human knowledge, has been the secret of western science since the seventeenth century, for it achieves a corporate collective power that is far greater than any one individual can exert. Primary scientific papers are not meant to be final statements of indisputable truths; each is merely a tiny tentative step forward, through the jungle of ignorance.

Nature
Volume 224, 1969 (p. 318)

It is not enough to observe, experiment, theorize, calculate and communicate; we must also argue, criticize, debate, expound, summarize, and otherwise transform the information that we have obtained individually into reliable, well established, public knowledge.

Nature
Volume 224, 1969 (p. 318)

There are two experiences on which our visual world is based: that gravity is vertical, and that the horizon stands at right angles to it.
Jacob Bronowski – (See p. 143)

CONCEPT

Taylor, E.S.
It is necessary to have a concept before it can be analyzed.

Journal of Engineering Education
Report on Engineering Design (p. 649)
Volume 51, Number 8, April 1961

CONSTRUCTION

Kipling, Rudyard
After me cometh a Builder. Tell him, I too have known.
When I was King and a Mason—a Master proven and skilled—
I cleared me ground for a Palace such as a King should build.
I decreed and dug down to my levels. Presently, under the silt,
I came on the wreck of a Palace such as a King had built . . .

⋮

Swift to my use in my trenches, where my well-planned ground-works
   grew,
I tumbled his quions and his ashlars, and cut and reset them anew.
Lime I milled of his marbles; burned it, slacked it, and spread;
Taking and leaving at pleasure the gifts of the humble dead.
Yet I despised not nor gloried; yet as we wrenched them apart,
I read in the razed foundation the heart of that builder's heart.
As he had risen and pleaded, so did I understand
The form of the dream he had followed in the face of the thing he had
   planned.

Collected Verse of Rudyard Kipling
The Palace (pp. 257–8)

CREATE

Shaw, George Bernard
You imagine what you desire; you will what you imagine; and you create what you will.

Back To Methuselah
Part I, Act I (p. 8)

Wilson, Robert Q.
He who is truly creative can distinguish between those matters worthy of change and those that are not worth the effort.

Battelle Technical Review
Volume 11, Number 4, April 1962 (p. 12)

CREATIVITY

Beveridge, W.I.B.
It is more important to have a clear understanding of general principles, without, however, thinking of them as fixed laws, than to load the mind with a mass of detailed technical information which can readily be found in reference books or card indexes.

The Art of Scientific Investigation
Preparation (p. 4)

Bolz, Ray E.
Dean, Robert C., Jr.
What the educational experience almost completely excludes is the exercise and development of the students' creativity—even though creativity is probably the single most important characteristic demanded of a modern, practicing engineer.

In Daniel V. DeSimone
Education for Innovation
Strategies and Teaching Methods
Chapter 11 (p. 128)

Freund, C.J.
For one thing, engineering creativity is much more like inventiveness than like research. The creative engineer is a cousin to Edison and Marconi; he is no relation at all to Einstein or Enrico Fermi.

Machine Design
Creativity is a Task, Not a Trait (p. 161)
25 May 1967

Gilmer, Ben S.
We need men who have been schooled in the principles of creativity and who dare to court the ridicule of the masses for the sake of improving the lot of mankind.

Auburn Alumnews
Times Demand as Goal: Education for Creativity (p. 7)
July 1961

Hoyle, Fred
The nation that neglects creative thought today will assuredly have its nose ground into the dust of tomorrow.

In Sidney J. Parnes and Harold F. Harding
A Source Book for Creative Thinking (p. 17)

King, Blake
Creativity bothers engineers.

Mechanical Engineering
Object: Creativity (p. 38)
November 1963

Moore, A.D.
Throughout the long history of the human race, we find that creativity has nearly always had to struggle against anything from discouragement to violent rejection.

Invention, Discovery, and Creativity (p. 140)

Morton, Jack A.
Creativity is involved in research, discovery of new knowledge, in its application, in development engineering, in the manufacture of the hardware, in marketing and sales, in the raising of capital, and in the supplying of services.

In Daniel V. DeSimone
Education for Innovation
Innovation and Entrepreneurship
Discussion (p. 105)

Tumin, Melvin
Let us not kid ourselves. The way to the creative life for the average man is difficult in the extreme.

In Sidney J. Parnes and Harold F. Harding
A Source Book for Creative Thinking (p. 113)

CREED

National Society of Professional Engineers
As a Professional Engineer, I dedicate my professional knowledge and skill to the advancement and betterment of human welfare.

I Pledge:
- To give the utmost of performance;
- To participate in none but honest enterprise;
- To live and work according to the laws of man and the highest standards of professional conduct;
- To place service before profit, the honor and standing of the profession before personal advantage, and the public welfare above all other considerations.

In humility and with need for Divine Guidance, I make this pledge.

Engineers' Creed
Adopted June 1954

I PLEDGE...

Unknown

Engineers have different objectives when it comes to social interaction.

"Normal" people expect to accomplish several unrealistic things from social interaction:

- Stimulating and thought-provoking conversation.
- Important social contacts.
- A feeling of connectedness with other humans.

In contrast to "normal" people, engineers have rational objectives for social interactions:

- Get it over with as soon as possible.
- Avoid getting invited to something unpleasant.
- Demonstrate mental superiority and mastery of all subjects.

Source unknown

I take the vision which comes from dreams and apply the magic of science and mathematics, adding the heritage of my profession and my knowledge of nature's materials to create a design.

I organize the efforts and skills of my fellow workers employing the capital of the thrifty and the products of many industries, and together we work toward our goal undaunted by hazards and obstacles.

And when we have completed our task all can see that the dreams and plans have materialized for the comfort and welfare of all.

I am an Engineer. I serve mankind by making dreams come true.

Engineers' Creed
Source unknown

DATA

Berkeley, Edmund C.

There is no substitute for honest, thorough, scientific effort to get correct data (no matter how much it clashes with preconceived ideas). There is no substitute for actually reaching a correct chain of reasoning. Poor data and good reasoning give poor results. Good data and poor reasoning give poor results. Poor data and poor reasoning give rotten results.

Computers and Automation
Right Answers—A Short Guide for Obtaining Them (p. 20)
September 1969

Deming, William Edwards

Anyone can easily misuse good data.

Some Theory of Sampling (p. 18)

There is only one kind of whiskey, but two broad classes of data, good and bad.

The American Statistician
On the Classification of Statistics (p. 16)
Volume 2, Number 2, April 1948

Scientific data are not taken for museum purposes; they are taken as a basis for doing something. If nothing is to be done with the data, then there is no use in collecting any. The ultimate purpose of taking data is to provide a basis for action or a recommendation for action. The step intermediate between the collection of data and the action is prediction.

Journal of the American Statistical Association
On a Classification of the Problems of Statistical Inference
Volume 37, June 1942

Hodnett, Edward

When you learn how to mobilize your data and bring them to bear on your problems, you are no longer a rank amateur.

The Art of Problem Solving (p. 42)

James, William
The ignoring of data is, in fact, the easiest and most popular mode of obtaining unity in one's thought.

Mind
The Sentiment of Rationality (p. 320)
Series 1, Volume 4, 1879

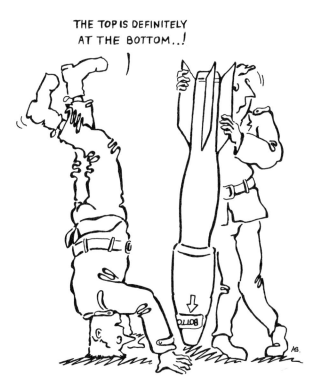

THE TOP IS DEFINITELY AT THE BOTTOM..!

It is necessary for technical reasons that these warheads be stored upside down, that is, with the top at the bottom and the bottom at the top...
British Admiralty – (See p. 44)

DECISION

British Admiralty
It is necessary for technical reasons that these warheads be stored upside down, that is, with the top at the bottom and the bottom at the top. In order that there may be no doubt as to which is the bottom and which is the top, it will be seen to it that the bottom of each warhead immediately be labeled with the word TOP.

Applied Optics
Of Optics and Opticists (p. 19)
Volume 7, Number 1, January 1968

Carroll, Lewis
It may be right to go ahead, I guess:
It may be right to stop, I do confess;
Also, it may be right to retrogress.

The Complete Works of Lewis Carroll
The Elections to the Hebdomadal Council

da Vinci, Leonardo
Think well to the end, consider the end first.

The Notebooks of Leonardo da Vinci
Volume I
Philosophy (p. 64)

Holmes, Sherlock
Take time to consider. The smallest point may be the most essential.

In Arthur Conan Doyle
The Complete Sherlock Holmes
The Adventure of the Red Circle

Huxley, Thomas H.
And when you cannot prove that people are wrong, but only that they are absurd, the best course is to let them alone.

Huxley's Essays
On the Method of Zadig (p. 171)

Popper, Karl R.
We do not know: we can only guess.

The Logic of Scientific Discovery (p. 278)

Proverb, Chinese
To guess is cheap, to guess wrongly is expensive.

Simon, Herbert
Decisions are something more than factual propositions. To be sure, they are descriptive of a future state of affairs, and this description can be true or false in a strictly empirical sense; but they possess, in addition, an imperative quality—they select one future state of affairs in preference to another and direct behavior toward the chosen alternative. In short, they have an *ethical* as well as factual content.

Administrative Behavior
Chapter III (p. 46)

Taylor, E.S.
. . . the most important decisions in a design problem must often be made without assistance from higher mathematics.

Journal of Engineering Education
Report on Engineering Design (p. 649)
Volume 51, Number 8, April 1961

Unknown
Judgment is needed to make important decisions on imperfect knowledge in limited time.

Source unknown

It is better to be inconsistent but sometimes right, than to be consistently wrong.

Source unknown

It's harder to change a decision than to make one.

Source unknown

DESIGN

Adams, Douglas
A common mistake that people make when trying to design something completely foolproof is to underestimate the ingenuity of complete fools.

Mostly Harmless
Chapter 12 (p. 135)

Layton, Edwin T., Jr.
From the point of view of modern science, design is nothing, but from the point of view of engineering, design is everything. It represents the purposive adaptation of means to reach a preconceived end, the very essence of engineering.

Technology and Culture
American Ideologies of Science and Engineering (p. 696)
Number 4, October 1976

Mailer, Norman
Indeed the early history of rocket design could be read as the simple desire to get the rocket to function long enough to give an opportunity to discover where the failure occurred. Most early debacles were so benighted that rocket engineers could have been forgiven for daubing the blood of a virgin goat on the orifice of the firing chamber.

Of a Fire on the Moon
Chapter I, section v (p. 165)

Reswick, J.B.
Design is the essential purpose of engineering.

In Morris Asimow
Introduction to Design
Foreword

Vincenti, Walter

Engineering knowledge reflects the fact that design does not take place for its own sake and in isolation. Artificial design is a social activity directed at a practical set of goals intended to serve human beings in some direct way. As such, it is intimately bound up with economic, military, social, personal, and environmental needs and constraints.

What Engineers Know and How They Know It (p. 11)

Designing a bridge or any other large structure is not unlike planning a trip or vacation. The end may be clear and simple: to go from here to there. But the means may be limited only by our imagination.
Henry Petroski – (See p. 19)

DISCOVERY

Bacon, Francis
Brutes by their natural instincts have produced many discoveries, whereas men by discussion and the conclusions of reason have given birth to few or none.

Novum Organum
Aphorisms, LXXIII

Bernard, Claude
A great discovery is a fact whose appearance in science gives rise to shining ideas, whose light dispels many obscurities and shows us new paths.

An Introduction to the Study of Experimental Medicine
Part I, Chapter II, section ii (p. 34)

. . . a discovery is generally an unforeseen relation not included in theory, for otherwise it would be foreseen.

An Introduction to the Study of Experimental Medicine
Part I, Chapter II, section iii (p. 38)

Ardent desire for knowledge, in fact, is the one motive attracting and supporting investigators in their efforts; and just this knowledge, really grasped and yet always flying away before them, becomes at once their sole torment and sole happiness. Those who do not know the torment of the unknown cannot have the joy of discovery, which is certainly the liveliest that any man can feel.

An Introduction to the Study of Experimental Medicine
Part III, Chapter IV, section iv (pp. 221–2)

Brougham, Henry

A discovery in mathematics, or a successful induction of facts, when once completed, cannot be too soon given to the world. But as an hypothesis is a work of fancy, useless in science, and fit only for the amusement of a vacant hour . . .

Edinburgh Review
Volume 1, January 1803 (p. 451)

De Costa, J. Chalmers

Many a man who is brooding over alleged mighty discoveries reminds me of a hen sitting on billiard balls.

The Trials and Triumphs of the Surgeon
Chapter I

de Maistre, J.

Those who have made the most discoveries in science are those who knew Bacon least, while those who have read and pondered him, like Bacon himself, have not succeeded well.

In H. Selye
From Dream to Discovery (p. 263)

Fischer, Martin H.

Every discovery in science is a tacit criticism of things as they are. That is why the wise man is invariably called a fool.

In Howard Fabing and Ray Marr
Fischerisms

Holton, G.

And yet, on looking into the history of science, one is overwhelmed by evidences that all too often there is no regular procedure, no logical system of discovery, no simple, continuous development. The process of discovery has been as varied as the temperament of the scientist.

The Scientific Imagination: Case Studies (pp. 384–5)

Mitchell, S. Weir

The success of a discovery depends upon the time of its appearance.

In F.H. Garrison
Bulletin of the New York Academy of Medicine
Volume 4, 1928 (p. 1002)

Moynihan, Berkeley, Sir

A discovery is rarely, if ever, a sudden achievement, nor is it the work of one man; a long series of observations, each in turn received in doubt and discussed in hostility, are familiarized by time, and lead at last to the gradual disclosure of truth.

Surgery, Gynecology & Obstetrics
Volume 31, 1920 (p. 549)

Rutherford, Ernest

It is not in the nature of things for any one man to make a sudden, violent discovery; science goes step by step and every man depends on the work of his predecessors. When you hear of a sudden unexpected discovery—a bolt from the blue, as it were—you can always be sure that it has grown up by the influence of one man or another, and it is the mutual influence which makes the enormous possibility of scientific advance. Scientists are not dependent on the ideas of a single man, but on the combined wisdom of thousands of men, all thinking of the same problem and each doing his little bit to add to the great structure of knowledge which is gradually being erected.

In Robert B. Heywood
The Works of the Mind
The Scientist (p. 178)

Safonov, V.

There are scientists who make their chief discovery at the threshold of their scientific career, and spend the rest of their lives substantiating and elaborating it, mapping out the details of their discovery, as it were. There are other scientists who have to tread a long, difficult and often tortuous path to its end before they succeed in crowning their efforts with a discovery.

Courage
Chapter 10 (p. 40)

Sigerist, Henry E.
We must also keep in mind that discoveries are usually not made by one man alone, but that many brains and many hands are needed before a discovery is made for which one man receives the credit.

A History of Medicine
Volume I
Introduction (p. 13)

Smith, Theobald
Great discoveries which give a new direction to currents of thoughts and research are not, as a rule, gained by the accumulation of vast quantities of figures and statistics. These are apt to stifle and asphyxiate and they usually follow rather than precede discovery. The great discoveries are due to the eruption of genius into a closely related field, and the transfer of the precious knowledge there found to his own domain.

Boston Medical and Surgical Journal
Volume 172, 1915 (p. 121)

Strauss, Maurice B.
Discoveries do not arise *de novo*, like Athene from the brow of Zeus, but are more akin to the living layers of a coral reef built on the past labors of countless predecessors.

Medicine
Volume 43, 1964 (p. 619)

Szent-Györgyi, Albert
Discovery consists of seeing what everybody has seen and thinking what nobody has thought.

In Irving John Good (Editor)
The Scientist Speculates
Chapter I, 6 (p. 15)

If there wasn't anything to find out, it would be dull. Even trying to find out and not finding out is just as interesting as trying to find out and finding out; and I don't know but more so.

Eve's Diary

ELECTRICAL

Unknown

1. Beware of lightning that lurketh in an uncharged condenser lest it cause thee to bounce upon thy buttocks in a most embarrassing manner.

2. Cause thou the switch that supplieth large quantities of juice to be opened and thusly tagged, that thy days may be long in this earthly vale of tears.

3. Prove to thyself that all circuits that radiateth, and upon which thou worketh, are grounded and thusly tagged lest they lift thee to a radio frequency potential and causeth thee to make like a radiator, too.

4. Tarry thou not amongst these fools that engage in intentional shocks for they are not long for this world and are surely unbelievers.

5. Take care that thou useth the proper method when thou takest the measures of high-voltage circuits, that thou dost not incinerate both thee and thy test meter, for verily, though thou has no company property number and can be easily replaced, the test meter has one and, as a consequence, the loss of which bringeth much woe unto a purchasing agent.

6. Take care that thou tamperest not with interlocks and safety devices, for this incurreth the wrath of the chief electrician and brings the fury of the engineers on his head.

7. Work thou not on energized equipment for if thou doest so, thy friends will surely be buying beers for thy widow and consoling her in certain ways not generally acceptable to thee.

8. Verily, verily I say unto thee, never service equipment alone, for electrical cooking is a slow process and thou might sizzle in thy own fat upon a hot circuit for hours on end before thy maker sees fit to end thy misery and drag thee into his fold.

9. Trifle thee not with radioactive tubes and substances lest thou commence to glow in the dark like a lightning bug, and thy wife be frustrated and have not further use for thee except for thy wages.

10. Commit thou to memory all the words of the prophets which are written down in thy Bible which is the National Electrical Code, and giveth out with the straight dope and consoleth thee when thou hast suffered a ream job by the chief electrician.

Electricians' Ten Commandments
Source unknown

ENERGY

Blake, William
Energy is the only life . . . and Reason is the bound or outward circumference of Energy.

The Prophetic Writings of William Blake
The Marriage of Heaven and Hell
The Voice of the Devil
#2

Energy is Eternal Delight.

The Prophetic Writings of William Blake
The Marriage of Heaven and Hell
The Voice of the Devil
#3

Dyson, Freeman J.
We do not know how the scientists of the next century will define energy or in what strange jargon they will discuss it. But no matter what language the physicists use they will not come into contradiction with Blake. Energy will remain in some sense the lord and giver of life, a reality transcending our mathematical descriptions. Its nature lies at the heart of the mystery of our existence as animate beings in an inanimate universe.

Scientific American
Energy in the Universe (p. 51)
Volume 224, Number 3, September 1971

Einstein, Albert
Some recent work by E. Fermi and L. Szilard . . . leads me to expect that the element uranium may be turned into a new and important source of energy in the immediate future.

The World As I See It
Letter of August 1939 to President Franklin D. Roosevelt

Emerson, Ralph Waldo
Coal is a portable climate. It carries the heat of the tropics to Labrador and the polar circle; and it is the means of transporting itself withersoever it is wanted. Watt and Stephenson whispered in the ear of mankind their secret, that *a half-ounce of coal will draw two tons a mile*, and coal carries coal, by rail and by boat, to make Canada as warm as Calcutta, and with its comfort brings its industrial power.

The Works of Ralph Waldo Emerson
Volume VI
Conduct of Life
Wealth

Esar, Evan
Atomic energy is the most powerful force known to man—except woman.

20,000 Quips and Quotes

Hammond, Allen Lee
Metz, William D.
A point about solar energy that government planners seem to have trouble grasping is that it is fundamentally different from other energy sources. Solar energy is democratic. It falls on everyone and can be put to use by individuals and small groups of people. The public enthusiasm for solar is perhaps as much a reflection for this unusual accessibility as it is a vote for the environmental kindliness and inherent renewability of energy from the sun.

Science
Solar Energy Research: Making Solar After the Nuclear Model? (p. 241)
Volume 197, Number 4300, 15 July 1977

Johnson, George
The weapons laboratory of Los Alamos stands as a reminder that our very power as pattern finders can work against us, that it is possible to discern enough of the universe's underlying order to tap energy so powerful that it can destroy its discoverers or slowly poison them with its waste.

Fire in the Mind (p. 326)

Lilienthal, David E.
Atomic energy bears the same duality that has faced man from time immemorial, a duality expressed in the Book of Books thousands of years ago: "See, I have set before thee this day life and good and death and evil . . . therefore choose life."

This I Do Believe (pp. 144–5)

Porter, T.C.
Old Dr. Joule he made this rule:
  The self-same energee
Which lifts a gram of matter to
  42640 c. (centimeters)
Will heat a gram of water through
  One centigrade degree.

Radio Times
February 1933

Siegel, Eli
Energy, like grammar, should be used correctly; the unjust expenditure of energy or its unjust withholding should cease immediately.

Damned Welcome (p. 38)

Teller, Edward
It took us eighteen months to build the first nuclear power generator; it now takes twelve years; that's progress.

In Milton Friedman and Rose Friedman
Free to Choose (p. 191)

ENGINEER

Alger, John R.M.
Hays, Carl V.
The engineer is concerned with creating material objectives to serve human needs . . . The engineer uses the knowledge and understanding developed by the scientist. In the absence of such knowledge, the engineer proceeds to a schedule by making judicious assumptions about the designs in order to insure successful solutions.

Creative Synthesis in Design (p. 2)

An inexperienced engineer commonly believes that he can accomplish about twice as much as he can in fact accomplish in a given time.

Creative Synthesis in Design (p. 58)

Alger, Philip
Christensen, N.A.
Olmsted, Sterling P.
It takes a wise man to give the right answer to a technical question that involves the conflicting rights and desires of a number of people. Yet the engineer is often required to give such an answer and on very short notice.

Ethical Problems in Engineering (p. 4)

The greatest difference between engineers and other professionals is that engineers do not offer a person-to-person service, nor are their services a matter of life, freedom, or property, of deep personal concern. Most engineers thrive on promotion and expansion of business, new inventions, new buildings, and more consumption of goods. Obviously, then, they are tempted to whip up trade, advertise, seek out clients, and promote deals. This spirit leads to competition. Until this is recognized, I see no hope of solving the problem of engineering professional development.

Ethical Problems in Engineering (p. 4)

Aronin, Ben
I'll carry the story along from here
And sing you the song of the engineer.

In Lenox Lohr
Centennial of Engineering: 1852–1952
Adam to Atom
Scene XV
Ballad of the Engineer—1840–1865 (p. 89)

Arwaker, Edmund
I learnt t'intrench a Camp, and Bulwarks rear,
With all the Cunning of an *Engineer*.

Pia Desidera
Book the Second
l. 805–806

Bailey, P.J.
Even as when
In planning some steel-rutted road, long years
Dreamed of,—where now the fire-horse ramps, steam-breath'd,
Sweating red coal-drops on his panting path,—
The deep-eyed engineer his level lays
Inscrutable, and anon, the hills with men,
Brood of his brain swarm; . . .

Festus
Chapter XXVIII, l. 17970 (p. 472)

Baillie, Joanna
Some thousand carcasses, living and dead,
Of those who first shall glut the en'my's rage,
Push'd in, pell-mell, by those who press behind
Will rear for us a bridge to mount the breach
Where ablest engineers had work'd in vain.

Miscellaneous Plays
Constantine Paleologus
l. 23–27

Birchmore, Sue
The best scientists are poets, the real engineer is an artist.

In E. Garfield
Current Comments
Creativity and Science
Part I (p. 296)
Number 43, 23 October 1989

Bishop, S.
Two Rooms, on one foundation set;
Mere walls, and floors, and ceilings yet:
But Taste, my Landlord's engineer,
Stands bound to finish 'em next year.

To the Reverend Mr. Fayting
l. 41–44

Blackie, John Stuart
Or know as engineer to guide
Blind forces with fine skill,
And bind the strong-breathed steam with bonds
To work thy reasoned will.

The Chief End of Man

Blough, Roger M.
We do not do ourselves a good turn by becoming panicky at the idea of
the mere number of engineers that are being produced in other countries,
or by consciously engaging in a technological numbers race.

American Engineer
Volume 26, Number 7, July 1956 (p. 5)

Boelter, L.M.K.
Engineers participate in the activities which make the resources of nature
available in a form beneficial to man and provide systems which will
perform optimally and economically.

In Ralph J. Smith
Engineering as a Career (p. 9)

Born, Max
The engineer, who has cunningly contrived to make the blind and deaf
molecules in their mad, senseless rush drive an engine, may well feel
proud of himself.

The Restless Universe (p. 16)

Bradley, Duane
[Engineers] are men with wings on their minds. These wings are courage
and imagination—when an engineer decides that a job is not to big for
him, and starts finding a way to do it, he is using his wings.

Engineers Did It! (p. 8)

Brandeis, Louis D.
The engineer spoke in figures—a language implying certitude.

In Alfred Lief (Editor)
The Social and Economic Views of Mr. Justice Brandeis (p. 141)

Brome, Alexander
All trades did show their skill in this,
Each wife an Engineer:
The Mayoress took the tool in hand,
The maids the stones did bear.

The Works of the English Poets
On Demolishing the Forts

Buchanan, Scott
Watch the engineer and you will learn many things, but do not ask him about mathematics, unless you want to see quite another thing, how technology and folk-lore get invented and broadcast.

Poetry and Mathematics
Poetry and Mathematics (p. 38)

The prestige of the engineer is another accretion to the tradition of mathematics. This more than any other one thing accounts for our present mathematical complex. The engineer is fast taking the position of authority, superseding the priest, the scholar, and the statesman in our organized thought and action.

Poetry and Mathematics
Poetry and Mathematics (pp. 37–8)

The engineer sings as he works—often he only whistles—and in that singing there is the magic of poetry. The engineer's science, like the sailor's chanty, is good literature.

Poetry and Mathematics
Poetry and Mathematics (pp. 38–9)

Cavendish, Margaret, Duchess of Newcastle
Envy and Malice were two Engineers,
Which Subtilty had Practis'd many Years:
Their Drum was Ignorance, Stupidity
Was one Stick, th'other was Obstinacy;
And Brac'd it was with Rudeness, which sounds Harsh
On Strings of Wilfulness that's ever Rash.

A Battle between Honour and Dishonour
l. 55–60

Churchill, Winston
We want a lot of engineers in the modern world, but we do not want a world of engineers. We want some scientists, but we must keep them in their proper place.

Speech
University of London, 18 November 1948
In F.B. Czarnomski
The Wisdom of Winston Churchill (p. 363)

Colclaser, R.G.
In many segments of industry, the young man is looked upon first as an *engineer* and second as an *electrical* engineer with specific knowledge. Industrial problems do not divide themselves neatly into the areas of electrical, mechanical, chemical, or civil engineering; rather, they are considered as general problems requiring a solution which the young engineer is expected to provide. In the majority of cases he must produce a physical device which will accomplish a desired result. This is not a "textbook" solution but an original, creative effort.

Engineering Education
A Design School for the Young Engineer in Industry (p. 812)
March 1968

Compton, Karl Taylor
If the engineer is to bring his influence to bear on broad public questions he must approach them, not with technical arrogance, but with sympathetic understanding.

The Journal of Engineering Education
Engineering and Social Progress (p. 14)
Volume 30, Number 1, September 1939

For the benefit of society, as well as for the most efficient work of the engineer, it is essential that the engineer should be trained to think not only of his specific engineering projects, but also of their larger significance in the economic and social order.

A Scientist Speaks (p. 49)

Cook, Morris L.
The more I think of it, the more I feel that the fundamental consideration in the work of an engineer—if he is ever to pull himself out of his present status of being a hired servant—is that he shall make public interest the master test of his work.

Letter to A.G. Christie, 9 June 1921
In Edwin T. Layton
The Revolt of the Engineers (p. 159)

Crichton, Michael
"Let's keep it in perspective," Hammond said. "You get the engineering correct and the animals will fall into place."

Jurassic Park
The Tour (p. 141)

Cross, Hardy
It is important that men know that engineers do not build alone with concrete and steel or by formulas and charts, but more than anything else by faith, hope and charity—faith in their methods, their training, in the

men with whom they work, faith in humanity, in the worth-whileness of life; hope that by use of these they may find men, money, materials and methods, not blind wishes but judicious hopes; charity that involves the sympathetic understanding of the human element and willingness.

Engineers and Ivory Towers
The Education of an Engineer (p. 59)

Darrow, Karl K.
But here is another who took up the study of physics as soon as he came to college, and continued it all through his student days, and his career consists in controlling and directing physical phenomena by his knowledge of physical laws, or in designing machines which depend on physical principles. And what does he call himself, and what does the world call him? An electrical engineer or a radio engineer, a designer of lenses or a maker of turbines, a naval engineer or an acoustical engineer, a mechanical engineer or an aerodynamical engineer, and only the census tables could say what else!

The Renaissance of Physics (p. 4)

Davis, Chandler
Any applied mathematicians—any engineer using mathematics—works sometimes more and sometimes less mathematically. When he is most mathematical he makes least appeal to experience.

Boston Studies in the Philosophy of Science
Volume XV
For Dirk Struk
Materialist Mathematics (p. 38)

de Beauvoir, Simone
He was living like an engineer in a mechanical world. No wonder he had become dry as a stone.

The Mandarins
Chapter 3 (p. 156)

de Camp, L. Sprague
Civilization, as we know it today, owes its existence to the engineers. These are the men who, down the long centuries, have learned to exploit the properties of matter and the sources of power for the benefit of mankind. By an organized, rational effort to use the material world around them, engineers devised the myriad comforts and conveniences that mark the difference between our lives and those of our fore-fathers thousands of years ago.

The story of civilization is, in a sense, the story of engineering—that long and arduous struggle to make the forces of nature work for man's good.

The Ancient Engineers (p. 13)

An engineer is merely a man who, by taking thought, tries to solve human problems involving matter and energy. Since the Mesopotamians tamed their first animal and planted their first seed, engineers have solved a multitude of such problems. In doing so, they have created the teeming, complex, gadget-filled world of today.

The Ancient Engineers (p. 372)

de Vega, Lope
Es amor grande ingeniero:
Las máquinas de Arquimedes
No son encarecimiento
Para las que tiene amor.

[Love is a mighty engineer,
Not Archimedes' skill could add
One jot or tittle to the power
Of the machines that Love controls.]

La Hermosa Fea
Act II, scene 7

Dean, Robert C., Jr.
. . . the job of the engineer is to change the world . . .

In Daniel V. DeSimone
Education for Innovation
Trade-Offs and Constraints (p. 111)

Defoe, D.
The Legislators are the Engineers,
Who when 'tis out of Order make Repairs:
The People are the Owners, 'twas for them
The first Inventer drew the antient Scheme.
'Tis for their Benefit it works, and they
The Charges of maintaining it defray:
And if their Governours unfaithful prove,
They, Engineers or Managers remove.
Unkind Contention sometimes there appears
Between the Managers and Engineers:
Such Strife is always to the Owners wrong,
And once it made the Work stand still too long;
Till William came, and loos'd the Fatal Chain,
And set the Engineers to work again:
And having made the wondrous thing compleat,
To Anne's unerring Hand he left the Helm of State.

Selected Poetry and Prose of Daniel Defoe
The Mock Mourners
l. 92–107

DeSimone, Daniel V.
. . . an engineer is supposed to be more than a mobile repository of knowledge who is adept at attacking single-answer problems.

Education for Innovation
Introduction (p. 13)

Dibdin, Charles Isaac Mungo
Goletta's walls, for tactic science fam'd,
Boast of the age, impregnable proclaim'd,
No dam afford to stop the raging tide,
All hearts conjoin'd, and ev'ry nerve applied;
The bulwark cracks; the engineer applies
Incessant art's destructive energies . . .

Young Arthur
Subject VI
Lament, l. 2738–2743

Dieudonné, Jean
Engineers, always looking for optimal values for the measures of magnitudes which interest them, think of mathematicians as custodians of a fund of formulae, to be supplied to them on demand.

Mathematics—The Music of Reason (p. 7)

Dodge, A.Y.
. . . for many engineers treat all new things pessimistically.

In Joseph Rossman
Industrial Creativity: The Psychology of the Inventor (p. 220)

Dumas, Hal S.
It is the engineer who must always be the link between the idea and actuality, between the probable and the practical. It is he who makes realities out of dreams. He is indeed the solvent which blends together the many different parts of our great industrial mechanism and produces a smoothly working whole.

In Lenox R. Lohr
Centennial of Engineering 1852–1952
The Telephone Engineer and His Job (p. 750)

It is the job of the engineer to search out the means by which the ideas of the inventor can be put to work in the service of the public. He also knocks on the door of the ivory tower of the pure scientist and calls forth new inventions to meet the needs and wants of the public.

In Lenox R. Lohr
Centennial of Engineering 1852–1952
The Telephone Engineer and His Job (p. 751)

Dunning, John R.
How engineers and scientists who daily drink thinly diluted, but treated sewage, and on a rationed basis at that; travel to and from work via arteries of congested, noisy traffic through asphalt jungles of soot and acid-blackened buildings; breathe smog-filled air symptomized by hacking and coughing, and tear-filled eyes, pay exorbitant clothing and home cleaning bills; and see the progressive deterioration of natural beauty; [how they can] continue to tolerate all of these things and more without doing something about it is incomprehensible.
The Urban Frontier and the Engineer
A pamphlet distributed by the Office of School Relations, Columbia University

Durand, William Frederick
. . . no one knows better than the engineer the need of discrimination between the sure ground of known data and formal logic, on the one hand—as exemplified, say, by mathematical operations—and acts of judgment on the other; and no one has learned through wider experience than the engineer the need of applying his conclusions in the light of that component part which, of necessity, has been dependent on estimate and judgment.
Transactions of the American Society of Mechanical Engineers
Presidential Address
Volume 47, 1925

Dyson, Freeman J.
A good scientist is a person with original ideas. A good engineer is a person who makes a design that works with as few original ideas as possible. There are no prima donnas in engineering.
Disturbing the Universe
Chapter 10 (p. 114)

Edwards, Llewellyn Nathaniel
Engineers bridging chasms wide
So people may reach the other side;
Dreamers, yes, but doers too,
Developing the strange and new.
A Record of History and Evolution of Early American Bridges
Builders of Bridges (p. xi)

The engineer historian gathers a meed of satisfaction and pleasure in tracing the progress made by predecessors in his art; reviewing their accomplishments; analyzing their solutions of problems; and examining the monuments of their industry and skill. He sees definitely and unmistakably that advances made in a given period have marked the way to greater accomplishments in a succeeding period.
A Record of History and Evolution of Early American Bridges (p. xii)

Egerton, Sarah
So some unlucky Engineer
Does all the fit Materials compound,
That are in Art or Nature found;
Will glorious Fire-Works prepare.

Poems on Several Occasions
The Advice
Part II, l. 41–44

Eisenhower, Dwight D.
Engineers build for the future, not merely for the needs of men but for their dreams as well. Thus, inherently, the engineer's work is a fearless optimism that life will go forward, and that the future is worth working for.

American Engineer
December 1951 (p. 5)

Emerson, Ralph Waldo
The machine unmakes the man. Now that the machine is so perfect, the engineer is nobody.

The Works of Ralph Waldo Emerson
Volume III
Society and Solitude: Work and Days (p. 67)

Emmet, William LeRoy
The all-important word to the engineer is WHY, and it is astonishing how few people in the ordinary pursuits of human affairs ever think it worthwhile to trouble themselves about that question, or to make much effort to find out whether the answer suggested will bear analysis.

The Autobiography of an Engineer (p. 226)

Evans, Sebastian
And there we kept pacing to and fro,
In a frenzy of mute surmising:
Quoth the Engineer in a whisper low:
"Is the tide in the river rising?"

A Tale of a Trumpeter
l. 61–64

"Parapet, buttress, and arch and pier
Beyond are as sound as ever!
Now show us thy skill, Sir Engineer,
For a roadway over the River!"

A Tale of a Trumpeter
l. 117–120

Farquhar, George
. . . rise thou prostrate Ingineer, not all thy undermining Skill shall reach my Heart.

The Beaux' Stratagem
Act V, scene II (p. 59)

Finch, James Kip
From the earliest days of recorded history the men now known as engineers—civil, chemical and mechanical, mining and metallurgical—have provided for man's unfolding material needs and wants. Engineering has, in fact, not only played an ever increasing role in man's material life, but has had a profound effect on human relations, in shaping modern social, economic, and political aims, ideals, and institutions. Indeed, the promises which engineering and engineered industry offer today stir the hopes, interest, and ambitions of the less-developed nations of the world.

The Story of Engineering (p. v)

Flanders, Ralph E.
There is an engineering approach to (our) problems. The engineer has an objective. He studies and analyzes the materials with which he has to deal. He acquaints himself with the natural forces which he cannot change, which are more powerful than he is, but to which he must adapt himself so that he may make use of them.

American Engineer
November 1952 (p. 6)

Florman, Samuel C.
The nation needs engineers who are able to communicate, who are prepared for leadership roles, who are sensitive to the worthy objectives of our civilization and the place of technology within it, and whose creative imaginations are nourished from the richest possible sources—spiritual, intellectual, and artistic. Furthermore, engineers as a group need to preserve their professional self-esteem—and the esteem of the greater community—by guarding against an insensitive mechanical approach to the work they do. And finally, individual engineers deserve the chance to enrich their lives with art, literature, history—the best our civilization has to offer.

The Civilized Engineer (p. 195)

The civil engineer, with his hands literally in the soil, is existentially wedded to the earth, more so than any other man except perhaps the farmer.

The Existential Pleasures of Engineering (pp. 121–2)

... public-safety policies are properly established, not by well-intentioned engineers, but by legislators, bureaucrats, judges, and juries, in response to facts presented by expert advisers ... It would be a poor policy indeed that relied upon the impulse of individual engineers.

Blaming Technology (p. 174)

Taken as a whole, engineers—and the technologists, craftsmen, and tinkerers from whom they spring—are, and always have been, as decent, moral, and law-abiding a group of men as one could find. Absorbed in their technical pursuits, they are singularly free of the greed, duplicity, and hostility that characterize so many of their fellows and that have caused so much grief.

Technology and Culture
Comment: Engineers and the End of Innocence (p. 14)
Volume 10, Number 1, January 1969

Freund, C.J.
People are not afraid of engineers, thank goodness, but they are excessively afraid of much that engineers contrive.

The Journal of Engineering Education
Engineering Education and Freedom from Fear (p. 11)
Volume 40, Number 1, September 1949

Surely there is no greater menace in the world than a superbly competent engineer who is equally content to engage himself to a benefactor of the human race or to some monster of cruelty and vice.

The Journal of Engineering Education
Engineering Education and Freedom from Fear (p. 11)
Volume 40, Number 1, September 1949

Freyssinet, E.
Some people will say that a respect for regulations is essential and that engineers need not check the hypotheses on which they are based. This is a convenient theory, but a false one. Men who draw up regulations can be wrong like other men.

The Birth of Prestressing

Gabor, Dennis
Short of a compulsory humanistic indoctrination of all scientists and engineers, with a 'Hippocratic oath' of never using their brains to kill people, I believe that the best makeshift solution at present is to give the alpha-minuses alternative outlets for their dangerous brain-power, and this may well be provided by space research.

Inventing the Future (p. 156)

Galbraith, John Kenneth
. . . the enemy of the market is not ideology but the engineer.

The New Industrial State
Chapter III, section 5 (p. 35)

Gillmor, R.E.
My observation . . . is that the engineer feels his professional responsibility to mankind just as much as does the physician and the military man. He does not (always) take an oath as they do, but his sense of professional responsibility is as deep in his heart as any oath could make it.

American Engineer
Volume 26, Number 9, September 1956 (p. 5)

Glegg, Gordon L.
A scientist can discover a new star but he cannot make one. He would have to ask an engineer to do it for him.

In Issac Asimov
Isaac Asimov's Book of Science and Nature Quotations (p. 79)

Golder, H.Q.
The scientist is interested in the right answer, the engineer in the best answer now.

In Robert F. Legget
Geology and Engineering
Foreword (p. xii)

Grace, Eugene G.
Thousands of engineers can design bridges, calculate strains and stresses, and draw up specifications for machines, but the great engineer is the man who can tell *whether* the bridge or the machine should be built at all, *where* it should be built, and *when*.

In George C. Beakley
Careers in Engineering and Technology (p. 33)

Gruenberg, Benjamin C.
Scientific studies develop . . . the habit of mind that submits every idea to rigid test. Much of the loose thinking in social, political, and economic affairs would be avoided if workers in these fields possessed real training in scientific thinking. The scientist and engineer have built the modern world, and they hold the key to its control and coordination.

Science and the Public Mind
Chapter IV (p. 34)

Hamerton, Philip Gilbert
Some have iron thews and sinews, some are muscular of mind;
Learned savans, skilful blacksmiths, each are noble in their kind.

But to give the savan's wisdom to the hammer and the shears,
Come those intermediate workers,—England's civil engineers.

The Britannia Bridge
l. 1–4

Herbert, George
Wit's an unruly engine, wildly striking
Sometimes a friend, sometimes the engineer.

The Country Parson, The Temple
The Church Porch, l. 241–242 (p. 129)

Hersey, John
I became an engineer.

A Single Pebble (p. 3)

Hogben, Lancelot
This is not the age of pamphleteers. It is the age of the engineers. The
spark gap is mightier than the pen. Democracy will not be salvaged by
men who can talk fluently, debate forcefully, and quote aptly.

Science for the Citizen
Epilogue (p. 1075)

Hood, Thomas
John Jones he was a builder's clerk,
On ninety pounds a year,
Before his head was engine-turn'd
To be an engineer!

The Comic Poems of Thomas Hood
John Jones, A Pathetic Ballad
l. 1–4

Hoover, Herbert
"I hope you will forgive my dreadful curiosity, but I should like awfully
to know—what is your profession?"

I replied that I was an engineer. She emitted an involuntary exclamation,
and said "Why, I thought you were a gentleman!"

The Memoirs of Herbert Hoover
Volume I
The Profession of Engineering (pp. 132)

It is a great profession. There is the satisfaction of watching a figment of the imagination emerge through the aid of science to a plan on paper. Then it moves to realization in stone or metal or energy. Then it brings jobs and homes to men. Then it elevates the standards of living and adds to the comforts of life. That is the engineer's high privilege. The great liability of the engineer compared to men of other professions is that his works are out in the open where all can see them. His acts, step by step, are in hard substance. He cannot bury his mistakes in the grave like the doctors. He cannot argue them into thin air or blame the judge like the lawyers. He cannot, like the architects, cover his failures with trees and vines. He cannot, like the politicians, screen his shortcomings by blaming his opponents and hope that the people will forget. The engineer simply cannot deny that he did it. If his works do not work, he is damned. That is the phantasmagoria that haunts his nights and dogs his days. He comes from the job at the end of the day resolved to calculate it again. He wakes in the night in a cold sweat and puts something on paper that looks silly in the morning. All day he shivers at the thought of the bugs which will inevitably appear to jolt his smooth consummation.

On the other hand, unlike the doctor his is not a life among the weak. Unlike the soldier, destruction is not his purpose. Unlike the lawyer, quarrels are not his daily bread. To the engineer falls the job of clothing the bare bones of science with life, comfort and hope. No doubt as years go by people forget which engineer did it, even if they ever knew. Or some politician puts his name on it. Or they credit it to some promoter who used other people's money with which to finance it. But the engineer himself looks back at the unending stream of goodness that flows from his successes with satisfactions that few professions may know. And the verdict of his fellow professionals is all the accolade he wants.

The Memoirs of Herbert Hoover
Volume I
The Profession of Engineering (pp. 132–3)

Howland, W.E.
Wiley, R.B.
It might be supposed that engineers, who are so largely responsible for the increase in the productive capacity of men and of nations, would be the first to enjoy, or at least to seek to enjoy, the benefits of their own accomplishments; that they would utilize, or seek to utilize in a prolonged period of education, the leisure time made possible by the improvement in productive efficiency which they brought about. That they do not do so is evidence of the need for the enrichment of their education: they don't know what they are missing.

Civil Engineering
Backsight at a Turning Point (pp. 199–200)
Volume XI, Number 4, April 1941

Huxley, Thomas H.

I ask any one who has adopted the calling of an engineer, how much time he lost when he left school, because he had to devote himself to pursuits which were absolutely novel and strange, and of which he had not obtained the remotest conception from his instructors.

Macmillian's Magazine
Scientific Education: Notes of an After Dinner Speech (p. 178)
Volume XX, July 1869

Johnson, Eric

Our present day diplomats are engineers and they take less pleasure from the marble fountain in a formal garden than from the sinking of a tube well or construction of an irrigation system.

American Engineer
September 1952 (p. 6)

Johnson, James Weldon

And so we ride
Over land and tide,
Without a thought of fear—
Man never had
The faith in God
That he has in an engineer!

Fifty Years & Other Poems
The Word of An Engineer

Keyser, Cassius J.

The characteristic marks of the great engineer will be four:
Magnanimity—Scientific Intelligence—Humanity—Action.

Mathematical Philosophy: A Study of Fate and Freedom
Science and Engineering (p. 462)

Killian, Dr. James R., Jr.

. . . you can be neither an effective scientist, engineer, executive, economist, nor architect without acquiring understanding of our society and of human relationships.

American Engineer
Volume 26, Number 5, May 1956 (p. 3)

Kingsley, Charles

So give me the political economist, the sanitary reformer, the engineer; and take your saints and virgins, relics and miracles. The spinning-jenny and the railroad, Cunard's liners and the electric telegraph, are to me . . . signs that we are, on some points at least, in harmony with the universe.

The Works of Charles Kingsley
Volume II
Yeast
Chapter 5 (p. 82)

Kipling, Rudyard
When the Waters were dried an' the Earth did appear,
('It's all one,' says the Sapper),
The Lord He created the Engineer,
Her Majesty's Royal Engineer,
With the rank and pay of a Sapper.

Collected Verse of Rudyard Kipling
Sappers

Krutch, Joseph Wood
Electronic calculators can solve problems which the man who made them cannot solve; but no government subsidized commission of engineers and physicists could create a worm.

The Twelve Seasons
March (p. 184)

Layton, Edwin T., Jr.
The engineer is both a scientist and a businessman. Engineering is a scientific profession, yet the test of the engineer's work lies not in the laboratory, but in the marketplace.

The Revolt of the Engineers (p. 1)

Engineers, as a rule are not and do not pretend to be philosophers in the sense of building up consistent systems of thought following logically from certain premises. If anything, they pride themselves on being hard-headed practical men concerned only with facts, disdaining mere speculation or opinion. In practice, however, engineers do make many assumptions about the nature of the universe, of man, and of society.

The Revolt of the Engineers (p. 53)

The cement binding the engineer to his profession was scientific knowledge. All the themes leading toward a closer identification of the engineer with his profession rested on the assumption that the engineer was an applied scientist. It was the cumulative character of scientific knowledge that gave weight to engineers' claims to be the agents of progress and enlightenment.

The Revolt of the Engineers (p. 58)

Le Corbusier [Jeanneret-Gris, Charles Édouard]
The Engineer, inspired by the law of Economy and governed by mathematical calculation, puts us in accord with universal law. He achieves harmony.

Towards a New Architecture (p. 7)

Engineers fabricate the tools of their time. Everything, that is to say, except houses and moth-eaten boudoirs.

Towards A New Architecture (p. 18)

Our engineers are healthy and virile, active and useful, balanced and happy in their work. Our architects are disillusioned and unemployed, boastful or peevish. This is because there will soon be nothing more for them to do. *We no longer have the money* to erect historical souvenirs. At the same time, we have got to wash!

Our builders provide for these things and they will be our builders.

Towards A New Architecture (pp. 18–19)

Our engineers produce architecture, for they employ a mathematical calculation which derives from natural law, and their works give us the feeling of HARMONY. The engineer therefore has his own aesthetic, for he must, in making his calculations, qualify some of the terms of his equation; and it is here that taste intervenes. Now, in handling a mathematical problem, a man is regarding it from a purely abstract point of view, and in such a state, his taste must follow a sure and certain path.

Towards A New Architecture (p. 19)

Lovejoy, Thomas E.
Natural species are the library from which genetic engineers can work.

Time
in Jamie Murphy
The Quiet Apocalypse (p. 80)
13 October 1986

Genetic engineers don't make new genes, they rearrange existing ones.

Time
in Jamie Murphy
The Quiet Apocalypse (p. 80)
13 October 1986

Mackay, Charles
Old King Coal was a merry old soul:
Quoth he, "We travel slow;
"I should like to roam the wide world round,
"As fast as the wild winds blow."
And he call'd for his skilful engineers;
And soon through hills and vales,
O'er rivers wide, through tunnels vast,
The flying trains like lightning pass'd,
On the ribs of the mighty Rails.

The Collected Songs
Old King Coal
l. 28–36

Mason, William
Here midway down, upon the nearer bank
Plant thy thick row of thorns, and, to defend
Their infant shoots, beneath, on oaken stakes,
Extend a rail of elm, securely arm'd
With spiculated pailing, in such sort
As, round some citadel, the engineer
Directs his sharp stoccade.

The English Garden: A Poem
Book the Second
l. 288–294

McCullough, David
Engineers who read, who paint, who grow roses and collect fossils
and write poetry, who fall asleep in lectures, very human-like, even
civilized civil engineers are scattered all through the historical record.
Civil engineers have been known to go to the theater, yes indeed; they
have taken pleasure in good music and fine wine and the company
of charming women. There is even historical evidence of the existence
among a few civil engineers of a sense of humor.

Civil Engineering
Civil Engineers Are People (p. 47)
December 1978

Michener, James A.
Scientists are men who dream about doing things. Engineers do them
. . . if you want to be an engineer but find you have ten thumbs, you
become a scientist.

Space (p. 235)

Mitchell, Margaret
The South produced statesmen and soldiers, planters and doctors,
lawyers and poets, but certainly not engineers or mechanics. Let Yankees
adopt such low callings.

Gone With the Wind

Morison, George S.
We are the priests of the new epoch without superstitions.

Transactions of the American Society of Civil Engineers
Address at the Annual Convention (p. 483)
June 1895

Any man who is thoroughly capable of understanding and handling a
machine may be called a mechanical engineer, but only he who knows
the principles behind that machine so thoroughly that he would be able

to design it or adapt it to a new purpose . . . can be classed as a civil engineer.

Transactions of the American Society of Civil Engineers
Address at the Annual Convention
June 1895

Morley, Christopher

Having made up our mind to become an engineer, we thought it would be a mistake not to take advantage of all possible aid.

The Powder of Sympathy
Adventures of a Curricular Engineer (p. 82)

Murray, Robert Fuller

They went to the north, they went to the south,
And into the west went they,
Till they found a civil, civil engineer,
And unto him did say:

'Now tell to us, thou civil engineer,
If this be fit to drink.'
And they showed him a cup of the town water,
Which was as black as ink.

Robert F. Murray: His Poems
A Ballad of the Town Water

Pagnol, Marcel

Il faut se méfier des ingénieurs: ça commence par la machine à coudre, ça finit par la bombe atomique.
[One has to look out for engineers: they begin with sewing machines and end up with the atomic bomb.]

Critique des critiques
Chapter 3 (p. 38)

Petroski, Henry

Engineers . . . are not superhuman. They make mistakes in their assumptions, in their calculations, in their conclusions. That they make mistakes is forgivable; that they catch them is imperative. Thus it is the essence of modern engineering not only to be able to check one's own work, but also to have one's work checked and to be able to check the work of others.

To Engineer Is Human (p. 52)

The work of the engineer is not unlike that of the writer. How the original design for a new bridge comes to be may involve as great a leap of the imagination as the first draft of a novel.

To Engineer is Human (p. 78)

The engineer no less than the poet sees the faults in his creations, and he learns more from his mistakes and those of others than he does from all the masterpieces created by himself and his peers.

To Engineer is Human (p. 82)

Pickering, William H.
We need a new kind of engineer, one who can build bridges to society as well as bridges across rivers.

The Bridge of ETA Kappa Gnu
The Engineer—1968 (p. 7)
May 1968

Rae, John A.
. . . the scientist wants to know chiefly for the sake of knowing; the engineer wants to know for the sake of using.

Technology and Culture
Science and Engineering in the History of Aviation (p. 291)
Fall 1961

Rankine, William John
Thus it is that the commonest objects are by science rendered precious; and in like manner the engineer or the mechanic, who plans and works with understanding of the natural laws that regulate the results of his operations, rises to the dignity of a Sage.

A Manual of Applied Mechanics (p. 11)

Rawnsley, Hardwick Drummond
Yet as I watch the marvelous engineer
Guess at wind-pressure, and on favouring wind
Send forth at will her silk from stores within,
One message for men's souls I seem to hear
"Let others live to eat, I eat to spin,
Joy's soul is work: God helps the worker's mind!"

The Spider's Message
At a Gilchrist Lecture
l. 9–14

Sand, George
Another places upon his nose a pair of paper or wooden spectacles; he performs the duty of engineer,—comes, goes, makes a plan, looks at the workmen, draws lines, plays the pedant, cries that everything is being ruined, causes the work to be abandoned and resumed at his will, and directs it at great length and as absurdly as possible.

The Haunted Pool
Chapter IV
The Cabbage

Sagan, Carl

"God" may be thought of as the cosmic watchmaker, the engineer who constructed the initial state and lit the fuse.

In Stephen Hawking
A Brief History of Time
Introduction (p. x)

Salisbury, J. Kenneth

One normally tends to catalog engineers either as analyzers or as synthesizers—the analyzers are the appraisers and evaluators; the synthesizers are those who are creative and ingenious in devising new ways of doing things. This sharp division is somewhat fallacious, however, because there is considerable overlapping.

General Electric Review
Qualities Industry Wants in Its Engineers (p. 17)
May 1952

Seeger, Peggy

When I was a little girl I wished I was a boy
I tagged along behind the gang and wore my corduroys.
Everybody said I only did it to annoy
But I was gonna be an engineer.

Recorded by Frankie Armstrong

Seely, Bruce E.

We have the man who fires the boiler and pulls the throttle dubbed a locomotive or stationary engineer; we have the woman who fires the stove and cooks the dinner dubbed the domestic engineer, and it will not be long before the barefooted African, who pounds the mud into the brick molds, will be calling himself a ceramic engineer.

Technology and Culture
SHOT, the History of Technology, and Engineering Education
Volume 36, Number 4, October 1995 (p. 744)

Shakespeare, William

For 'tis the sport to have the enginer
Hoist with his own petar.

Hamlet
Act III, scene 4, l. 206–207

Shaw, George Bernard

Very nice sort of place, Oxford, I should think, for people that like that sort of place. They teach you to be a gentleman there. In the Polytechnic they teach you to be an engineer or such like.

Man and Superman
Act II (p. 50)

Now there is no calculation that an engineer can make as to the behavior of a girder under a strain, of an astronomer as to the recurrence of a comet, more certain than the calculation that under such circumstances we shall be dismembered unnecessarily in all directions by surgeons who believe the operations to be necessary solely because they want to perform them.

The Doctor's Dilemma
Preface On Doctors (p. vi)

Shute, Nevil
It has been said that an engineer is a man who can do for ten shillings what any fool can do for a pound . . .

Slide Rule
Chapter 3 (p. 64)

Snow, C.P.
. . . engineers have to live their lives in an organized community, and however odd they are underneath they manage to present a disciplined face to the world.

Two Cultures and the Scientific Revolution (p. 33)

Pure scientists have by and large been dim-witted about engineers and applied science. They couldn't get interested. They wouldn't recognize that many of the problems were as intellectually exacting as pure problems, and that many of the solutions were as satisfying and beautiful. Their instinct—perhaps sharpened in this country by the passion to find a new snobbism wherever possible, and to invent one if it doesn't exist— was to take it for granted that applied science was an occupation for second rate minds. I say this more sharply because thirty years ago I took precisely that line myself.

The Two Cultures: and A Second Look (p. 32)

Stalin, Joseph
The writer is an engineer of the human soul.

In John Gunther
Inside Russia Today (p. 284)

Starkey, W.L.
The engineer knows that it is easier to analyze a machine than it is to design one. Engineering analysis is simpler than engineering synthesis or design.

Mechanical Engineering
The Ingredients of Design
May 1966

Stassen, Harold E.
The building of a just and durable peace absolutely requires the sustained strength which flows in such a large measure from the work of engineers.
American Engineer
Volume 26, Number 3, March 1956 (p. 3)

Sterrett, The Right Reverend Frank W.
Many a battle has been lost because men lacked confidence in the outcome. That has not been characteristic of the Engineer. He is accustomed to face hard tasks demanding his best. The rebuilding and restoring of an ordered world present such a problem. It seems to me there is a continuing place of dignity for the Engineer of tomorrow.
American Engineer
June 1951 (p. 3)

Stevenson, Robert Louis
The engineer of to-day is confronted with a library of acquired results; tables and formulae to the value of folios full have been calculated and recorded; and the student finds everywhere in front of him the footprints of the pioneers. In the eighteenth century the field was largely unexplored; the engineer must read with his own eyes the face of nature.
Records of a Family of Engineers
Chapter I (p. 212)

The seas into which his labours carried the new engineer were still scarce charted, the coasts still dark; his way on shore was often far beyond the convenience of any road; the isles in which he must sojourn were still partly savage. He must toss much in boats; he must often adventure on horseback by the dubious bridle-track through unfrequented wildernesses; he must sometimes plant his lighthouse in the very camp of wreckers; and he was continually enforced to the vicissitudes of outdoor life.
Records of a Family of Engineers
Chapter I (p. 213)

The engineer was not only exposed to the hazards of the sea; he must often ford his way by land to remote and scarce accessible places, beyond reach of the mail or the post-chaise, even the tracery of the bridle-path, and guided by natives across bog and heather.
Records of a Family of Engineers
Chapter II, Part I (p. 241)

Even the mechanical engineer comes at last to an end of his figures, and must stand up, a practical man, face to face with the discrepancies of nature and the hiatuses of theory.
Records of a Family of Engineers
Chapter II, Part III (p. 261)

With the civil engineer, more properly so called (if anything can be proper with this awkward coinage), the obligation starts with the beginning. He is always the practical man. The rains, the winds and the waves, the complexity and the fitfulness of nature, are always before him. He has to deal with the unpredictable, with those forces (in Smeaton's phrase) that 'are subject to no calculation'; and still he must predict, still calculate them, at his peril.

Records of a Family of Engineers
Chapter II, Part III (p. 261)

The duty of the engineer is twofold—to design the work, and to see the work done.

Records of a Family of Engineers
Chapter II, Part III (p. 265)

Thring, M.W.
. . . this type of engineer is the leaven that leavens the whole of the country. If a country has plenty of creative engineers doing real creative work, it moves forward with the times. If it does not, it falls behind, however good all its other people are.

Proceedings of the Institution of Mechanical Engineers
On the Threshold (p. 1089)
Volume 179, Part I, 1963–64

The creative engineer uses his three brains. He uses his intellectual brain as an applied scientist to understand the laws of science and to see that the things which he invents do not break these laws. He uses his emotional brain for the act of invention, and he uses his physical brain— his brain with his hands, feet and body—for the proper understanding of the design of things that will work.

Proceedings of the Institution of Mechanical Engineers
On the Threshold (p. 1089)
Volume 179, Part I, 1963–64

Unknown
When considering the behavior of a howitzer:
A mathematician will be able to calculate where the shell will land.
A physicist will be able to explain how the shell gets there.
An engineer will stand there and try to catch it.

Source unknown

Come and listen to a story bout a man named Jed,
A poor College Kid barely kept his family fed,
But then one day he was talking to a recruiter,
He said "They'll pay ya big bucks if ya work on a computer",
Unix that is . . . CRT's . . . Workstations;

Well the first thing ya know ol' Jed's an Engineer,
The kinfolk said "Jed move away from here",
They said "Arizona is the place ya oughta be",
So he bought some donuts and moved to Ahwatukee,
Intel that is . . . no ordinary company.

On his first day at work they stuck him in a cube,
Fed him more donuts and sat him at a tube,
They said "Your project's late but we know just what to do,
Instead of 40 hours, we'll work you fifty-two!"
Overtime that is, unpaid mind you.

The weeks rolled by and things were looking bad,
Some schedules slipped and some managers were mad,
They called another meeting and decided on a fix,
The answer was simple, "We'll work him sixty-six"
"O brother" said Jed, "I'm tired of the kicks".

Months turned into years and his hair was turning gray,
Jed worked hard while his life slipped away,
Waiting to retire when he turned sixty-four,
Instead he got a call and they escorted him out the door,
Laid-off that is . . . and ol' Jed couldn't take it no more.

Under the tree lies our Jed,
"As an engineer you'll be happy" they'd said,
Happy he is, but only when he's dead!

The Engineer's Song
(Sung To The Tune Of The Beverly Hillbillies)
Source unknown

A Programmer and an Engineer were sitting next to each other on an airplane. The Programmer leans over to the Engineer and asks if he wants to play a fun game. The Engineer just wants to sleep so he politely declines, turns away and tries to sleep. The Programmer persists and explains that it's a real easy game. He explains, "I ask a question and if you don't know the answer you pay me $5. Then you ask a question and if I don't know the answer I'll pay you $5." Again the Engineer politely declines and tries to sleep.

The Programmer, now somewhat agitated, says, "O.K., if you don't know the answer you pay me $5 and if I don't know the answer I pay you $50!" Now, that got the Engineer's attention, so he agrees to the game. The Programmer asks the first question, "What's the distance from the earth to the moon?" The Engineer doesn't say a word and just hands the Programmer $5.

Now, its the Engineer's turn. He asks the Programmer, "What goes up a hill with three legs and comes down on four?" The Programmer looks at him with a puzzled look, takes out his laptop computer, looks through all his references and after about an hour wakes the Engineer and hands the Engineer $50. The Engineer politely takes the $50, turns away and tries to return to sleep.

The Programmer, a little miffed, asks, "Well what's the answer to the question?" Without a word, the Engineer reaches into his wallet, hands $5 to the Programmer, turns away and returns to sleep.

Source unknown

In some foreign country a priest, a lawyer and an engineer are about to be guillotined. The priest puts his head on the block, they pull the rope and nothing happens—he declares that he's been saved by divine intervention—so he's let go. The lawyer is put on the block, and again the rope doesn't release the blade, he claims he can't be executed twice for the same crime and he is set free too. They grab the engineer and shove his head into the guillotine, he looks up at the release mechanism and says, "Wait a minute, I see your problem."

Source unknown

Chorus:
We are, we are, we are, we are the Engineers,
We can, we can, we can, we can demolish forty beers.
So come, so come, so come along with us
For we don't give a damn for any damn man who don't give a damn
for us.

An Artsman and an Engineer once found a gallon can.
Said the Artsman, "Match me drink for drink, as long as you can stand."
They drank three drinks, the Artsman fell, his face was turning green,
But the Engineer drank on and said "It's only gasoline."

Chorus

Venus is a statue made entirely of stone,
There's not a fig leaf on her, she's as naked as a bone.
On noticing her arms were broke an Engineer discoursed:
"Why, the damn thing's busted concrete and it should be reinforced!"

Chorus

My father peddles opium, my mother's on the dole,
My sister used to walk the streets, but now she's on parole.
My brother runs a restaurant with bedrooms in the rear,
But they don't even speak to me 'cause I'm an Engineer.

Chorus

On Friday nights I go down to the bars of ill repute
On Saturdays, I'm out again—Tequila do I shoot.
I pissed my parents off and they threw me on my rear,
But I don't really have a care 'cause I'm an Engineer.

Chorus

A drunken Engineer once staggered in through Roddick Gate.
He stumbled through the lecture hall so drunk and very late,
The only thing that held him up and kept him on his course,
Were the boundary conditions and the electromotive force.

Chorus

Sir Francis Drake and his men set out for Calais Bay,
They'd heard the Spanish Rum Fleet was headed up that way.
But the Engineers had them beat by a night and half the day,
And although as drunk as they could be, you still could hear them say:

Chorus

The army and the navy boys went out to have some fun,
Looking for the taverns where the fiery liquors run.
All they found were empties for the Engineers had come,
And traded in their instruments for a gallon keg of rum.

Chorus

Caesar set out for Egypt at the age of fifty three,
But Cleopatra's blood was red, her heart was warm and free.
And every night when Caesar said goodnight at three o'clock,
A Roman Engineer was waiting just around the block.

Chorus

Elvis was a legend: he's the King of Rock 'n Roll.
But the life that he was leading—well, it finally took its toll.
Then one day he realised he had chose the wrong career,
So he faked his death, joined us here—now he's an Engineer!

Chorus

On reading Kama Sutra, a guy learned position nine
For proving masculinity, it truly was divine.
But then one day, a girl rebelled and threw him on his rear
For he was a feeble artsie and she, an Engineer!

Chorus

Now you've heard our story and you know we're Engineers,
We love to love our women/men and we love to drink our beers.
We drink to every one who comes from far and near.
'Cause we're each a HELL-OF-A, HELL-OF-A, HELL-OF-AN
    ENGINEER!

MIT Engineer Drinking Song
http://www.mudcat.org

One day a young man was walking down a road when a frog called to him: "Boy, if you kiss me, I will turn into a beautiful princess."

The young man picked up the frog, smiled at it and put it in his pocket.

A short while later, the frog said, "Boy, if you kiss me and turn me back into a beautiful princess, I'll be yours."

The young man took the frog from his pocket, smiled at it and put it back.

Now the frog was upset. "Boy, what is the matter?" the frog cried. "I have told you that I am a beautiful princess, and if you kiss me, I'll be yours!"

The young man took the frog from his pocket, looked at it and said: "Look, I'm an engineer. I have no time for a girlfriend, but a talking frog is cool!"

Source unknown

Verily, I say unto ye,
marry not an engineer.
For an engineer is a strange being
and possessed of many evils.

Yea, he speaketh always in parables
which he calleth formulae.
He wieldeth a big stick
which he calleth a slide rule.
And he hath only one bible,
a handbook.

He thinketh only of strains and stresses,
and without end of thermodynamics.
He showeth always a serious aspect
and seemeth not to know how to smile.
He picketh his seat in a car by the springs thereof
and not by the damsels.

Neither does he know a waterfall
except by its horsepower,
Nor a sunset
except that he must turn on the light,
Nor a damsel
except by her weight.

Always he carrieth his books with him,
and he entertaineth his sweetheart with steam tables.
Verily, though his damsel expecteth chocolates when he calleth,
She openeth the package to discover samples of iron ore.

Yea, he holdeth her hand
but to measure the friction thereof,
and kisseth her
only to test the viscosity of her lips,
for in his eyes shineth a far away look
that is neither love nor longing,
but a vain attempt to recall formulae.

Even as a boy, he pulleth a girl's hair
but to test its elasticity.
But as a man,
he deviseth different devices.
For he counteth the vibrations of her heartstrings
And seeketh ever to pursue his scientific investigations.
Even his own heart flutterings
he counteth as a measure of fluctuation.

And his marriage is but a
simultaneous equation involving two unknowns.
And yielding diverse results.

Verily, I say unto ye,
do not marry an engineer.

Ann Landers Friday 11-19-93
Psalm to an Engineer's Sweetheart

A number of different approaches are being tried
We don't know where we're going, but we're moving.

Close project coordination
We should have asked someone else; or, let's spread the responsibility for this.

Customer satisfaction is believed assured
We are so far behind schedule that the customer was happy to get anything at all from us.

Developed after years of intensive research
It was discovered by accident.

Extensive effort is being applied on a fresh approach to the problem
We just hired three new guys; we'll let them kick it around for a while.

Major Technological Breakthrough
Back to the drawing board.

Modifications are underway to correct certain minor difficulties
We threw the whole thing out and are starting from scratch.

Preliminary operational tests are inconclusive
The darn thing blew up when we threw the switch

Project slightly behind original schedule due to unforeseen difficulties
We are working on something else.

Test results were extremely gratifying
It works, and are we surprised!

The designs are well within allowable limits
We just made it, stretching a point or two.

The design will be finalized in the next reporting period
We haven't started this job yet, but we've got to say something.

The entire concept will have to be abandoned
The only guy who understood the thing quit.

The Engineer's Dictionary
Source unknown

What is the difference between Mechanical Engineers and Civil Engineers?

Mechanical Engineers build weapons; Civil Engineers build targets.

Source unknown

There are four engineers traveling in a car; a mechanical engineer, a chemical engineer, an electrical engineer and a computer engineer.

The car breaks down.

"Sounds to me as if the pistons have seized. We'll have to strip down the engine before we can get the car working again", says the mechanical engineer.

"Well", says the chemical engineer, "it sounded to me as if the fuel might be contaminated. I think we should clear out the fuel system."

"I thought it might be a grounding problem", says the electrical engineer, "or maybe a faulty plug lead."

They all turn to the computer engineer who has said nothing and say: "Well, what to you think?"

"Ummm—perhaps if we all get out of the car and get back in again?"

Source unknown

To the engineer, all matter in the universe can be placed into one of two categories: (1) things that need to be fixed, and (2) things that will need to be fixed after you've had a few minutes to play with them. Engineers like to solve problems. If there are no problems handily available, they will create their own problems. Normal people don't understand this concept; they believe that if it ain't broke, don't fix it. Engineers believe that if it ain't broke, it doesn't have enough features yet.

Source unknown

Clothes are the lowest priority for an engineer, assuming the basic thresholds for temperature and decency have been satisfied. If no appendages are freezing or sticking together, and if no genitalia or mammary glands are swinging around in plain view, then the objective of clothing has been met. Anything else is a waste.

Source unknown

Engineers are always honest in matters of technology and human relationships. That's why it's a good idea to keep engineers away from customers, romantic interests, and other people who can't handle the truth.

Source unknown

The only distinction between physicists and engineers is the physicists have more questions than answers while engineers have more answers than questions.

Source unknown

A mathematician, a physicist, and an engineer were all given a red rubber ball and told to find the volume. The mathematician carefully measured the diameter and evaluated a triple integral. The physicist filled a beaker with water, put the ball in the water, and measured the total displacement. The engineer looked up the model and serial numbers in his red-rubber-ball table.

Source unknown

Engineers think that equations approximate the real world.
Physicists think that the real world approximates equations.
Mathematicians are unable to make the connection.

Source unknown

To be consistent the civil engineer should now vote for the erection and constant maintenance in New York of a hotel large enough to accommodate their entire membership in attendance at any convention to be held there and not to be used by any other persons at any other time, especially by persons who may be members of any other engineering organization or otherwise masquerading as engineers.

American Machinist
The Carnegie Union Engineering Building (p. 335)
Volume 27, 10 March 1904

Heaven is where
The police are British
The cooks are French
The engineers are German

The administrators are Swiss
And the lovers are Italian

Hell is where
The police are German
The cooks are British
the engineers are Italian
The administrators are French
And the lovers are Swiss

Letter in *The Independent*
2 June 1990

. . . whenever an engineer learns something new in technics, it is his bounden duty to put it in writing and see that it is published where it will reach the eyes of his confrères and be always available to them. It is absolutely a crime for any man to die possessed of useful knowledge in which nobody else shares.

In J.A.L. Waddell, Frank W. Skinner and Harold E. Wessman (Editors)
Vocational Guidance in Engineering Lines
First Edition (p. 10)

[Chemical engineers] . . . we are not even able to persuade the engineers that we are engineers.

American Institute of Chemical Engineers Bulletin
Number 24, 1921 (p. 53)

A good engineer must be of inflexible integrity, sober, truthful, accurate, resolute, discrete, of cool and sound judgment; must have courage to resist and repeal attempts at intimidation, a firmness that is proof against solicitation, flattery, or improper bias of any kind; must take an interest in his work; must be energetic, quick to decide, prompt to act; must be as fair and impartial as a judge on the bench; must have experience in his work and in dealing with men, which implies some maturity of years; and must have business habits and knowledge of accounts. Men who combine these qualities are not to be picked up every day. Still they can be found, and when found, they are worth their price; rather they are beyond price, and their value cannot be estimated by dollars.

In J.A.L. Waddell, Frank W. Skinner and Harold E. Wessman (Editors)
Vocational Guidance in Engineering Lines (p. 483)

Director of Admissions, my ma's an engineer,
My sister's a computer, my brother is a gear,
My father is a robot, my grandpa's also queer,
Leaping logic, naturally I'm here!

http://www.mudcat.org
Director of Admissions

When I was just a lad of ten
My father said to me:
Try hard to be an engineer
And get a PhD
Don't put your faith in Yale my boy,
My father said to me.
They'll teach you to make missiles
If you go to MIT.

http://www.mudcat.org
Lemon Tree (song)

Marching past, straight through to hell
The infantry are seen,
Accompanied by the Engineers,
Artillery and Marine,
For none but the shades of Cavalrymen
Dismount at Fiddler's Green.

http://www.mudcat.org
Fiddler's Green (song)

US Army Corps of Engineers
Essayon's
[Let's try!]

Motto

Vollmer, James
No longer can an engineer expect to work in a given specialty for most
of his life. Within five years a problem area of broad interest can be
completely mined out partly because of the number of miners, and partly
because of the sophistication of their equipment.

The Bridge of Eta Kappa Nu
Engineering, Growing, Steady State, or Evanescent
Volume 65, Number 4, August 1969

von Karman, Theodore
The scientist merely explores that which exists, while the engineer creates
what has never existed before.

Machine Design
Creativity is a Task, Not a Trait
25 May 1967

Walker, Eric A.
. . . when an engineer goes to work, he is no longer just an analyst of
problems but a synthesizer.

Engineering Education
Our Tradition-Bound Colleges (p. 89)
October 1969

. . . the modern engineer's primary concern should be that of designing and creating the things that society needs, and the spark of genius must be nurtured and developed to the maximum extent.

Held 21–25 June 1965
Report of the World Congress on Engineering Education
Engineering Education Around the World

Ward, Edward
Their Engineer his utmost Cunning try'd,
But found no Skreen could his Approaches hide;
For all the various Stratagems he us'd,
Ended thro' Royal Conduct, still confus'd.

A Fair Shell, BUT A Rotten Kernel, l. 233–236

Whittier, John
Beat by hot hail, and wet with bloody rain,
The myriad-handed pioneer may pour,
And the wild West with the roused North combine
And heave the engineer of evil with his mine.

The Complete Poetical Works of John Greenleaf Whittier
To a Southern Statesman

Wickenden, W.E.
. . . man-made machines and the harnessing of natural resources are progressively relieving humanity from the distress of an oppressively heavy physical toil and are affording improved opportunity for the development of mind and spirit. This is the challenging opportunity— and responsibility—of the engineer and his profession.

American Engineer
November 1951 (p. 7)

Wigner, Eugene
Part of the art and skill of the engineer and of the experimental physicist is to create conditions in which certain events are sure to occur.

Symmetries and Reflections
The Role of Invariance Principles in Natural Philosophy (p. 29)

Winne, Harry A.
It is the engineer's responsibility to take the new research discoveries as they come along and to put them to work for the benefit of man, and to find ways of doing it that industry and the people can afford.

American Engineer
17–23 February 1952 (p. 4)

Winsor, Dorothy A.
Engineers tend to prefer saying that they are being *convincing* rather than persuasive, and the very fact that they choose a different term suggests

that, at least for them, *persuasion* has associations that are not applicable to the relationship between engineers and their readers.

Writing Like an Engineer (p. 3)

Scientists and engineers may all study physical reality, but the scientist is usually considered successful if he or she has contributed to theory while the engineer is less interested in generating theory for its own sake than in doing whatever is necessary to design and produce useful artifacts. A scientist who does not understand a phenomenon has failed; but an engineer who does not fully understand a device may still be considered successful if the device works well enough. Scientists and engineers thus operate with different standards for success that affect the way they argue.

Writing Like an Engineer (p. 10)

Woodhouse, J.
Would engineer from steam of tea-pot-spout,
Expect to move his ponderous works about?

Ridicule, l. 315–316

Woodson, Thomas T.
. . . estimation and order-of-magnitude analysis are the hallmarks of the engineer.

Introduction to Engineering Design (p. 107)

The engineer uses all the analysis and quantification he can command; but in the end, the decisions are made *subjectively*; and there is no avoiding it.

Introduction to Engineering Design (p. 204)

Wright, Harold Bell
It was the last night out. Supper was over and the men, with their pipes and cigarettes, settled themselves in various careless attitudes of repose after the long day . . . All were strong, clean-cut, vigorous specimens of intelligent, healthy manhood, for in all the professions, not excepting the army and navy, there can be found no finer body of men than our civil engineers.

The Winning of Barbara Worth (pp. 86–7)

Zener, C.
Engineers have traditionally been people who work toward the attainment of practical goals. If a particular task requires the use of some practical goals. If a particular task requires the use of some particular physical phenomenon, then the more he understands this particular phenomenon the better able he will be to reach his goal. However, as

an engineer he could not care less about his understanding *per se*. In contrast, scientists have traditionally been people whose sole drive was to understand the world around them. They could not care less what use was made of this understanding.

Florida Engineer
Engineering in the Future
October 1965

HAVE YOU
FINISHED
WINDING
THAT CLOCK
UP YET?...

To the engineer, all matter in the universe can be placed into one of two categories: (1) things that need to be fixed, and (2) things that will need to be fixed after you've had a few minutes to play with them.
Unknown – (See p. 87)

ENGINEERING

American Society for Engineering Education
The engineering profession is the channel by which science can greatly improve our way of life, provided it assumes the initiative of leadership rather than the passive role of the hired consultant.

Goals of Engineering Education, The Preliminary Report (p. 15)

Asimov, Isaac
Science can amuse and fascinate us all, but it is engineering that changes the world.

Isaac Asimov's Book of Science and Nature Quotations (p. 78)

Billington, David
Engineering or technology is the making of things that did not previously exist, whereas science is the discovering of things that have long existed. Technological results are forms that exist only because people want to make them, whereas scientific results are formulations of what exists independently of human intentions.

The Tower and the Bridge (p. 9)

Science and engineering may share the same techniques of discovery—physical experiments, mathematical formulation—but students quickly learn that the techniques have vastly different applications in the two disciplines. Engineering analysis is a matter of observing and testing the actual working of bridges, automobiles, and other objects made by people, while scientific analysis relies on closely controlled laboratory experiments or observations of natural phenomena and on general mathematical theories that explain them. The engineer studies objects in order to change them; the scientist, to explain them.

Wilson Quarterly
The Defense of Engineers (p. 89)
Volume 10, Number 1, 1986

British Engineer to the Royal Aeronautical Society
Aeroplanes are not designed by science, but by art in spite of some pretence and humbug to the contrary. I do not mean to suggest that

engineering can do without science, on the contrary, it stands on scientific foundations, but there is a big gap between scientific research and the engineering product which has to be bridged by the art of the engineer.

In Walter G. Vincenti
What Engineers Know and How They Know It (p. 4)

Brown, Gordon Spencer
Doing engineering is practicing the art of the organized forcing of technological change.

Electronics
Engineer–Scientist (p. 53)
Volume 32, Number 47, 20 November 1959

Capp, Al
If half the engineering effort and public interest that go into the research of the American bosom had gone into our guided-missile program, we would now be running hot-dog stands on the moon.

Reader's Digest
July 1958

Compton, Karl Taylor
The characteristic feature of our age results from the wedding of science and engineering. It is the working together of disciplined curiosity and purposeful ingenuity to create new materials, new forces, and new opportunities which powerfully affect our manner of living and ways of thinking.

A Scientist Speaks (p. 1)

The question of engineering should be of interest not only to those of us who are engineers, but to the entire public which lives in an engineering world.

A Scientist Speaks (p. 49)

Engineering education is the *sine qua non* of this technical age. Unless it is effective and adequate our type of civilization cannot go forward. To be effective, it must be progressive, for engineering art is not static; it is very dynamic.

A Scientist Speaks (p. 49)

More recently in the development of a program of biological engineering, based upon physical, chemical, and biological operations, a similar attempt has been made to synthesize an appropriate training for the handling of a great variety of biological situations, whether they be in the food industry or in the hospital or medical or biological research fields. I suspect that there may be other directions in which an analogous approach may be made to simplify the educational program and at the same time increase the power acquired by the student.

A Scientist Speaks (p. 53)

Crick, Francis
Almost all aspects of life are engineered at the molecular level, and without understanding molecules we can only have a very sketchy understanding of life itself.

What Mad Pursuit
Chapter 5 (p. 61)

Cross, Hardy
It is customary to think of engineering as a part of a trilogy, pure science, applied science and engineering. It needs emphasis that this trilogy is only one of a triad of trilogies into which engineering fits. The first is pure science, applied science, engineering; the second is economic theory, finance and engineering; and the third is social relations, industrial relations, engineering. Many engineering problems are as closely allied to social problems as they are to pure science.

Engineers and Ivory Towers
The Education of an Engineer (p. 56)

Engineering then is not merely a mathematical science. It must be approached with a sense of proportion and aesthetics.

Engineering and Ivory Towers
The Education of an Engineer (p. 64)

DeSimone, Daniel V.
Engineering is a profession, an art of action and synthesis and not simply a body of knowledge. Its highest calling is to invent and innovate.

Education for Innovation
Introduction (pp. 1–2)

Elgerd, Olle I.
Engineering is an art of simplification, and the rules—when and how to simplify—are a matter of experience and intuition.

In Robert L. Bailey
Disciplined Creativity for Engineers
Table 14–1 (p. 433)

Emmerson, G.S.
Modern scientific principle has been drawn from the investigation of natural laws, technology has developed from the experience of doing, and the two have been combined by means of mathematical system to form what we call engineering.

Engineering Education: A Social History (p. 7)

Emmet, William LeRoy
Engineering is to a very large extent dependent upon detail . . .

The Autobiography of an Engineer (p. 225)

Everitt, W.L.
It is easier to distinguish between the 'scientific function' and the 'engineering function' than to distinguish between the man who should be called a scientist and who should be termed an engineer. Many men perform both functions, and do it very well . . .

In Panel on Engineering Infrastructure
Engineering Infrastructure Diagramming and Modeling (p. 73)

The Federated American Engineering Society
Engineering is the science of controlling the forces and of utilizing the materials of nature for the benefit of man, and the art of organizing and of directing human activities in connection therewith.

Preamble to Constitution (1920)

Ferguson, Eugene S.
Engineering drawings are expressed in a graphic language, the grammar and syntax of which are learned through use; it also has idioms that only initiates will recognize. And because the drawings are neatly made and produced on large sheets of paper, they exhume an air of great authority and definitive completeness.

Engineering and the Mind's Eye (p. 3)

Fish, J.C.L.
Every engineering structure, with few exceptions, is first suggested by economic requirements; and the design of every part, excepting few, and of the whole is finally judged from the economic standpoint.

Engineering Economics: First Principles (p. v)

Flinn, Alfred D.
In fact "engineering" now often signifies a new system of thought, a fresh method of attack upon the world's problems; the antithesis of traditionalism, with its precedents and dogmas.

Civil Engineering
Leadership in Economic Progress (p. 242)
Volume 2, Number 4, April 1932

Florman, Samuel C.
As engineering becomes increasingly central to the shaping of society, it is ever more important that engineers become introspective. Rather than merely revel in our technical successes, we should intensify our efforts to explore, define, and improve the philosophical foundations of our professions.

The Civilized Engineer (p. xii)

. . . without imagination, heightened awareness, moral sense, and some reference to the general culture, the engineering experience becomes less meaningful, less fulfilling than it should be.

The Civilized Engineer (p. 24)

Engineering is committed to the prospect of new discoveries, and engineers still look eagerly to ever receding horizons. We are tinkerers at heart; we cannot keep our hands off the world. However, the over-optimism, and perhaps even arrogance, that had been creeping into the engineering view is being replaced by a more thoughtful but still enthusiastic commitment to change.

The Civilized Engineer (p. 76)

When engineers attempt to write creatively . . . the results are usually disastrous.

Engineering and the Liberal Arts (p. 92)

Engineering is superficial only to those who view it superficially. At the heart of engineering lies existential joy.

The Existential Pleasures of Engineering (p. 101)

Friedel, Robert
Technology is simply not . . . largely the province of engineers, and "engineering" is certainly not coextensive with "technology".

Technology and Culture
Engineering in the 20th Century (p. 669)
Special issue, October 1968

Fung, Y.C.B.
. . . for engineering, the method is scientific, the mode is quantitative, the dictum is economy, the concern is human.

In David L. Arm
Journeys in Science
An Approach to Bioengineering (p. 108)

Gillette, H.P.
Engineering is the conscious application of science to the problems of economic production.

In Ralph J. Smith
Engineering as a Career (p. 8)

Grinter, L.E.
Engineering is far from static, for it is essentially a creative profession.

Journal of Engineering Education
Report on Evaluation of Engineering Education (1952–1955)
Volume 46, Number 1, September 1955

Hamilton, L.L.
It has been my lot for many years
To read reports by engineers:
Of projects, jobs, and tasks and such
From which I should learn very much.

Alas! alack! I am undid
And maybe soon I'll flip my lid,
Unless someone can help me out
And tell me what they're all about.

They do not mean what they say
But start in the middle and go each way.
The outline is missing, the form is lax
No useful punctuation, or correct syntax.

The object is a secret beautifully kept,
And conclusions avoided in manner adept.
Oh here I go, I've blown my top
About a thing no one can stop.
Just let me say before I'm muzzled,
I'm not the only one who's puzzled.

American Engineer
Engineer Report (p. 4)
Volume 29, Number 6, June 1959

Hammond, H.P.
Throughout the whole fabric of engineering education, therefore, there must be interwoven with instruction in scientific principles . . . the encouragement of creative talent . . .

Journal of Engineering Education
Engineering Education After the War (p. 599)
Volume 43, Number 9, May 1944

. . . the engineering profession clearly cannot isolate itself from this complex of men and functions as a well-defined caste . . .

Journal of Engineering Education
Report of Committee on Aims and Scope of Engineering Curricula
Volume 30, Number 7, March 1940

Hammond, John Hays
Chemical engineering more than any other, may be called the engineering of the future. It is the result of an evolution in which most of the other branches have played a part . . . The chemical engineer stands today on the threshold of a vast virgin realm; in it lie the secrets of life and prosperity for mankind in the future of the world.

In Lenox R. Lohr
Centennial of Engineering 1852–1952
The Story of Chemical Engineering (p. 176)

Harris, A.J.

The foundation of engineering is knowledge of materials, not, as engineers are so often apt to preach, a knowledge of mathematics; . . . knowledge of what [materials] are made of, how they are made, how they are shaped, how you fit them together, how they stand up to stress, how they break, how they catch fire, how they react to all the various agencies of ruin which are perpetually nibbling at them, how in due course they fall down.

Journal of the Royal Institute of British Architects
Architectural Misconceptions of Engineering (p. 130)
3rd Series, Volume 68

Hellmund, R.E.

Engineering is an activity other than pure manual and physical work which brings about the utilization of the materials and laws of nature for the good of humanity.

In Ralph J. Smith
Engineering as a Career (p. 8)

Holloman, J. Herbert

. . . we cannot effectively talk about the needs of engineering until we have reflected on the needs of society.

In Daniel V. DeSimone
Education for Innovation
Creative Engineering and the Needs of Society (p. 23)

There are deeply held feelings that engineering education has become too science-based and has become removed to some degree from the creative act that the engineer or inventor has to perform to bring the results of science and technology to the benefit of society.

In Daniel V. DeSimone
Education for Innovation
Creative Engineering and the Needs of Society (p. 23)

Hoover, Herbert

. . . engineering without imagination sinks to a trade.

The Memoirs of Herbert Hoover
Volume I
The Profession of Engineering (p. 132)

From the point of view of accuracy and intellectual honesty the more men of engineering background who become public officials, the better for representative government.

The Memoirs of Herbert Hoover
Volume I
The Profession of Engineering (p. 133)

Hoover, T.J.
Fish, J.C.L.
Engineering is the professional and systematic application of science to the efficient utilization of natural resources to produce wealth.

In Ralph J. Smith
Engineering as a Career (p. 8)

Huxley, Aldous
Some day, conceivably, the scientific and logical engineers may build us convenient bridges from one world to another. Meanwhile we must be content to hop.

Music at Night and Other Essays (p. 44)

Jeans, Sir James Hopwood
. . . we may say that we have already considered with disfavor the possibility of the universe having been planned by a biologist or an engineer; from the intrinsic evidence of his creation, the Great Architect of the Universe now begins to appear as a pure mathematician.

The Mysterious Universe
Into the Deep Waters (p. 165)

Kiddle, Alfred W.
Engineering is the art or science of utilizing, directing or instructing others in the utilization of the principles, forces, properties and substances of nature in the production, manufacture, construction, operation and use of things . . . or of means, methods, machines, devices and structures . . .

In Ralph J. Smith
Engineering as a Career (p. 7)

Kipling, Rudyard
"Good morrn, M'Andrew! Back again? An' how's your bilge to-day?"
Miscallin' technicalities but handin' me my chair
To drink Madeira wi' three Earls—the auld Fleet Engineer
That started as a boiler-whelp—when steam and he were low.

Collected Verse of Rudyard Kipling
McAndrew's Hymn (p. 35)

Kirkpatrick, Sidney D.
Chemical engineering is that branch of engineering concerned with the development and application of manufacturing processes in which chemical or certain physical changes of material are involved. These processes may usually be resolved into a coordinated series of unit physical operations and unit chemical processes. The work of the chemical engineer is concerned primarily with the design, construction

and operation of equipment and plants in which these unit operations and processes are applied. Chemistry, physics and mathematics are the underlying sciences of chemical engineering, and economics its guide in practice.

American Institute of Chemical Engineers Transactions
In Albert B. Newman
Development of Chemical Engineering Education in the United States (pp. 6–7)
Volume 34, Number 3a, 25 July 1938

Lawson, Henry Hertzberg

Ah, well! but the case seems hopeless, and the pen might
    write in vain;
The people gabble of old things, over and over again.
For the sake of the sleek importer we slave with the
    pick and shears,
While hundreds of boys in Australia long to be engineers.

Australian Engineer

Lindsay, S.E.

Engineering is the practice of safe and economic application of the scientific laws governing the forces and materials of nature by means of organization, design and construction, for the general benefit of mankind.

In Ralph J. Smith
Engineering as a Career (p. 8)

Little, Arthur D.

Chemical engineering as a science . . . is not a composite of chemistry and mechanical and civil engineering, but a science of itself, the basis of which is those unit operations which in their proper sequence and coordination constitute a chemical process as conducted on the industrial scale. These operations, as grinding, extracting, roasting, crystallizing, distilling, air-drying, separating, and so on, are not the subject matter of chemistry as such nor of mechanical engineering. Their treatment in the quantitative way . . . and . . . the materials and equipment concerned in them is the province of chemical engineering. It is this selective emphasis on the unit operations themselves in their quantitative aspects that differentiates chemical engineering from industrial chemistry, which is concerned primarily with general processes and products.

In Terry S. Reynolds
Technology and Culture
Defining Professional Boundaries: Chemical Engineering
in the Early 20th Century (p. 709)
Special issue, October 1968

Lower, Lennie

Talk of iron! We knew a man who had so much iron that he was full of nuts and bolts. Matter of fact, he lived on nuts and bolted his meals.

After he was operated on for appendicitis he had to be riveted. If he wanted to turn around, he had to use a spanner. Threw himself under a train and wrecked the train. Rusted away after a long and peaceful life, and was pronounced dead by one of the best engineers in the country.

Here's Another
Lonely Sardine (p. 102)

Macaulay, Thomas Babington

The mathematician can easily demonstrate that a certain power, applied by means of a certain lever, or of a certain system of pulleys, will suffice to raise a certain weight. But his demonstration proceeds on the sup-position that the machinery is such as no load will bend or break. If the engineer who has to lift a great mass of real granite by the instru-mentality of real timber and real hemp, should absolutely rely on the propositions which he finds in treatises on Dynamics and should make no allowance for the imperfection of his materials, his whole apparatus of beams, wheels, and ropes, would soon come down in ruin, and with all his geometrical skill, he would be found a far inferior builder to those painted barbarians who, though they never heard of the parallelogram of forces, managed to pile up Stonehenge.

In William John Rankine
A Manual of Applied Mechanics
Preliminary Dissertation (pp. 4–5)

Macdonald, Douglas D.

1. Information necessitating a change in design will be conveyed to designer only after the plans are complete.
2. The more innocuous the revision appears at first, the farther its influence will extend with time.
3. If, when the completion of a design is imminent, field dimensions are finally supplied as they actually are instead of what they were meant to be, it is always simpler to start over.
4. Even if it is impossible to assemble a part incorrectly, a way will be found to do it wrong.

In Harold Faber
The Book of Laws
Four Laws of Engineering and Design (p. 94)

Mailer, Norman

Physics was love and engineering was marriage.

Of a Fire on the Moon
Chapter I, section v (p. 165)

Physics was sex, conception and the communion of the family— engineering was getting the eggs out on time.

Of a Fire on the Moon
Chapter I, section v (p. 165)

McCune, Francis K.
. . . at the very core of engineering there is just one thing—an act of creative thought, or in other words the process of having an idea.

Elements of Competitive Engineering; 1965 Engineering Deans' symposium
Pamphlet by General Electric Co., nd

The characteristics of a productive facility and the signals from a social system furnish very specific facts which must become every bit as much a part of an engineering idea as any technology or scientific principle . . .

Elements of Competitive Engineering; 1965 Engineering Deans' symposium
Pamphlet by General Electric Co., nd

Engineering's prime mission is the creation of technical things and services useful to man.

Elements of Competitive Engineering; 1965 Engineering Deans' symposium
Pamphlet by General Electric Co., nd

Morison, George S.
Accurate engineering knowledge must succeed commercial guesses.

Transactions of the American Society of Civil Engineers
Address at the Annual Convention (p. 474)
June 1895

O'Brien, M.P.
The activity characteristic of professional engineering is the design of structures, machines, circuits, or processes, or of combinations of these elements into systems or plants and the analysis and prediction of their performance and costs under specified working conditions.

In Ralph J. Smith
Engineering as a Career (p. 8)

Parsons, William Barclay
It is not the technical excellence of an engineering design which alone determines its merit but rather the completeness with which it meets the economic and social needs of its day.

Address at the inauguration of the Columbia Student Chapter
of the American Society of Civil Engineers, 1927
In James Kip Finch
Engineering and Western Civilization (p. vii)

Petroski, Henry
Engineering, like poetry, is an attempt to approach perfection. And engineers, like poets, are seldom completely satisfied with their creations. They notice, even if no one else does, the word that is not quite *le mot juste* or the hairline crack that blemishes the structure.

To Engineer is Human (p. 83)

Piaget, Jean
Engineering—is more pragmatic and less—relatively less—speculative; it is production, operation, and management, as well as research, analysis, planning and design. Its goal is usually a clearly specified utility such as public health, communication, power, transportation, or housing, rather than the attainment of abstract truth.

In Thomas C. Dean
Phi Kappa Phi Journal
Challenges in Higher Education
Volume L, Number 3, Summer 1970

Rogers, G.F.C.
Engineering refers to the practice of organizing the design and construction [and, I would add, operation] of any artifice which transforms the physical world around us to meet some recognized need.

The Nature of Engineering, A Philosophy of Technology (p. 51)

Scott, Chas F.
Engineering is a mode of thinking.

Engineering Education
The Aims of the Society (p. 103)
Volume 12, Number 3, November 1921

Shewhart, W.A.
The fundamental difference between engineering with and without statistics boils down to the difference between the use of a scientific method based upon the concept of laws of nature that do not allow for chance or uncertainty and a scientific method based upon the concepts of laws of probability as an attribute of nature.

University of Pennsylvania Bicentennial Conference

Smith, Ronald B.
Engineering is the art of skillful approximation; the practice of gamesmanship in the highest form. In the end it is a method broad enough to tame the unknown, a means of combining disciplined judgment with intuition, courage with responsibility, and scientific competence within the practical aspects of time, of cost, and of talent.

This is the exciting view of modern-day engineering that a vigorous profession can insist be the theme for education and training of its youth. It is an outlook that generates its strength and its grandeur not in the discovery of facts but in their application; not in receiving, but in giving. It is an outlook that requires many tools of science and the ability to manipulate them intelligently.

In the end, it is a welding of theory and practice to build an early, strong, and useful result. Except as a valuable discipline of the mind, a formal education in technology is sterile until it is applied.

Mechanical Engineering
Professional Responsibility of Engineering (p. 18)
Volume 86, Number 1, January 1964

Engineering is a method and a philosophy for coping with that which is uncertain at the earliest possible moment and to the ultimate service to mankind. It is not a science struggling for a place in the sun. Engineering is extrapolation from existing knowledge rather than interpolation between known points. Because engineering is science in action—the practice of decision making at the earliest moment—it has been defined as the art of skillful approximation. No situation in engineering is simple enough to be solved precisely, and none worth evaluating is solved exactly. Never are there sufficient facts, sufficient time, or sufficient money for an exact solution, for if by chance there were, the answer would be of academic and not economic interest to society. These are the circumstances that make engineering so vital and so creative.

Mechanical Engineering
Engineering is . . .
Volume 86, Number 5, May 1964

Smith, Willard A.
Engineering is the science of economy, of conserving the energy, kinetic and potential, provided and stored up by nature for the use of man. It is the business of engineering to utilize this energy to the best advantage, so that there may be the least possible waste.

In Ralph J. Smith
Engineering as a Career (pp. 6–7)

Solzhenitsyn, Aleksandr
There are all kinds of engineers. Some of them here have built successful careers selling soda water.

The First Circle
Chapter 4 (p. 15)

Sporn, Philip
The scientist usually works—but very seldom under the pressure of a timetable—in a field of his special interest, in which he has generally chosen to stake out a narrow sector for his own specialization. The engineer, on the other hand, while also operating within the area of his own competence, has to tackle a variety of problems, some of which may be new to him, but to which he has to apply his scientifically based knowledge and skill to produce workable and practical solutions;

this work includes economics and involves both analysis and synthesis, generally within a rigid time limit. This is technology and engineering.

Foundations of Engineering (pp. 12–13)

Steinmetz, Charles Proteus

Indeed, the most important part of engineering work—and also of other scientific work—is the determination of the method of attacking the problem, whatever it may be.

In John Charles Lounsbury Fish
The Engineering Method (p. 1)

Engineering investigations evidently are of no value, unless they can be communicated to those to whom they are of interest.

In John Charles Lounsbury Fish
The Engineering Method (p. 290)

Stott, Henry G.

Engineering is the art of organizing and directing men and controlling the forces and materials of nature for the benefit of the human race.

In Ralph J. Smith
Engineering as a Career (p. 6)

Taylor, E.S.

The analytical part of an engineering education now seems to be considered the most difficult, most challenging part, while the remainder of engineering is considered to be an exercise of a lower order, conducted in a physical region located nearer the seat of the pants than the brain.

Journal of Engineering Education
Report on Engineering Design (p. 655)
Volume 51, Number 8, April 1961

Thring, M.W.

The roots of the tree are pure science . . . and a peculiar thing comes in here, called aesthetics, about which the architects and some other people are very concerned. The trunk of the tree is called human understanding and, in particular, applied mathematics. The branches of the tree are engineering and the extreme twigs are the growing new fields of engineering in which things are really happening . . . pure science is the roots which feed the tree but the actual growth of new life comes on the twigs of extremely specialized engineering. Some sort of scheme of knowledge like this is important.

Proceedings of the Institution of Mechanical Engineers
On the Threshold (pp. 1089, 1091)
Volume 179, Part I, 1964–65

Tredgold, Thomas
[Engineering] The art of directing the great sources of power in nature for the use and convenience of man, as the means of production and of traffic in states, both for external and internal trade, as applied in the construction of roads, bridges, harbours, moles, breakwaters, and lighthouses, and in the art of navigation by artificial power for the purposes of commerce, and in the drainage of cities and towns.

Charter 1828
Institution of Civil Engineers

Trefethen, Joseph M.
The relationship between civil engineering and geology is as old as the hills, man made hills that is . . .

Journal of Engineering Education
Geology for Civil Engineers (p. 383)
Volume 39, Number 7, March 1949

Unknown
Structural engineering is the science and art of designing and making, with economy and elegance, buildings, bridges, frameworks, and other similar structures so that they can safely resist the forces to which they may be subjected.

The Structural Engineer
Contents page

Every rocket-firing that is successful is hailed as a scientific achievement; every one that is a failure is regarded as an engineering failure.

Source unknown

Three engineering students were gathered together discussing the possible designers of the human body.

One said, "It was a mechanical engineer. Just look at all the joints."

Another said, "No, it was an electrical engineer. The nervous system has many thousands of electrical connections."

The last said, "Actually it was a civil engineer. Who else would run a toxic waste pipeline through a recreational area?"

Source unknown

Waddell, J.A.L.
Skinner, Frank W.
Wessman, H.E.
Engineering is more closely akin to the arts than perhaps any other of the professions; first, because it requires the maximum of natural aptitude and of liking for the work in order to offset other factors; second, because

it demands, like the arts, an almost selfless consecration to the job; and, third, because out of the hundreds who faithfully devote themselves to the task, only a few are destined to receive any significant reward—in either money or fame.

Vocational Guidance in Engineering Lines
Foreword (p. VI)

Engineering is the science and art of efficient dealing with materials and forces . . . it involves the most economic design and execution . . . assuring, when properly performed, the most advantageous combination of accuracy, safety, durability, speed, simplicity, efficiency, and economy possible for the conditions of design and service.

Vocational Guidance in Engineering Lines
Chapter II (p. 6)

Walker, Eric A.
Science aims at the discovery, verification, and organization of fact and information . . . engineering is fundamentally committed to the translation of scientific facts and information to concrete machines, structures, materials, processes, and the like that can be used by men.

Journal of Engineering Education
Engineers and/or Scientists (pp. 419–421)
Volume 51, February 1961

Wellington, Arthur Mellen
. . . to define it rudely but not inaptly, [engineering] is the art of doing that well with one dollar, which any bungler can do with two after a fashion.

The Economic Theory of the Location of Railways
Introduction (p. 1)

Winsor, Dorothy A.
Engineering is knowledge work. That is, although the goal of engineering may be to produce useful objects, engineers do not construct such objects themselves. Rather they aim to generate knowledge that will allow such objects to be built.

Writing Like an Engineer (p. 5)

EQUATION

Dirac, Paul Adrien Maurice
I consider that I understand an equation when I can predict the properties of its solutions, without actually solving it.

In Frank Wilczek and Betsy Devine
Longing for the Harmonies (p. 102)

It is more important to have beauty in one's equations that to have them fit experiment.

Scientific American
The Evolution of the Physicist's Picture of Nature
May 1963

Hertz, Heinrich
Maxwell's theory is Maxwell's system of equations.

Electric Waves (p. 21)

Thoreau, Henry David
I do believe in simplicity. When the mathematician would solve a difficult problem, he first frees the equation from all encumbrances, and reduces it to its simplest terms.

The Cambridge History of American Literature in Four Volumes
Book II, Chapter X
Thoreau (p. 8)

ERROR

Adams, Franklin Pierce
If frequently I fret and fume,
And absolutely will not smile,
I err in company with Hume,
Old Socrates and T. Carlyle.

Tobogganing on Parnassus
Erring in Company

Amiel, Henri Frédéric
An error is the more dangerous in proportion to the degree of truth which it contains.

Journal Intime (p. 43)

No forms of error are so erroneous as those that have the appearance without the reality of mathematical precision.

Journal Intime (p. 751)

Beard, George M.
. . . as quantitative truth is of all forms of truth the most absolute and satisfying, so quantitative error is of all forms of error the most complete and illusory.

Popular Science Monthly
Experiments with Living Human Beings (p. 751)
Volume 14, 1879

Borel, Émile
The problem of error has preoccupied philosophers since the earliest antiquity. According to the subtle remark made by a famous Greek philosopher, the man who makes a mistake is twice ignorant, for he does not know the correct answer, and he does not know that he does not know it.

Probability and Certainty
Chapter 9 (p. 114)

Cage, John Milton Jr.

An error is simply a failure to adjust immediately from a preconception to an actuality.

Silence 1961
45′ for a Speaker

Cicero

By Hercules! I prefer to err with Plato, whom I know how much you value, than to be right in the company of such men.

Tusculanarum Disputationum
I, 17

Colton, Charles Caleb

It is almost as difficult to make a man unlearn his errors, as his knowledge. Mal-information is more hopeless than non-information; for error is always more busy than ignorance. Ignorance is a blank sheet, on which we may write; but error is a scribbled one, from which we must first erase. Ignorance is contented to *stand still* with her back to the truth; but error is more presumptuous, and *proceeds* in the same direction. Ignorance has no light, but error follows a false one. The consequence is, that error, when she retraces her footsteps, has farther to go, before we can arrive at the truth, than ignorance.

Lacon (p. 17)

Diamond, Solomon

Error does not carry any recognizable badge, for when we change our point of view, to focus on a different problem, what had been error may become information, and what had been information may become error.

Information and Error (p. 7)

Here, by the grace of Chance, we've staked a Mean,
Uncertain marker of elusive Truth.
But have we caught a fact, or trapped a doubt
Within this stretching span of confidence—
A shadow world four standard errors wide,
All swollen by the stint of observation?
For recollect that once in twenty times
The phantom Truth will even lie beyond
That span, in the unending thin-drawn tails
Which point to the infinitude of Error.

Information and Error (p. 120)

Dryden, John

Errors, like straws, upon the surface flow,
He who would search for pearls must dive below.

The Poetical Works of Dryden
All for Love, Prologue, l. 25–26

Goethe, Johann Wolfgang von

Es giebt Menschen die gar nicht irren, weil sie sich nichts vernünftiges vorsetzen.

[There are men who never err, because they never propose anything rational.]

Spruche in Prosa
III

Heinlein, Robert A.

I shot an error into the air. It's still going . . . everywhere.

Expanded Universe (p. 514)

Helmholtz, Hermann von

But any pride I might have felt in my conclusions was perceptibly lessened by the fact that I knew that the solution of these problems had almost always come to me as the gradual generalisation of favourable examples, by a series of fortunate conjectures, after many errors. I am fain to compare myself with a wanderer on the mountains who, not knowing the path, climbs slowly and painfully upwards and often has to retrace his steps because he can go no further—then, whether by taking thought or from luck, discovers a new track that leads him on a little till at length when he reaches the summit he finds to his shame that there is a royal road, by which he might have ascended, had he only had the wits to find the right approach to it. In my works, I naturally said nothing about my mistake to the reader, but only described the made track by which he may now reach the same heights without difficulty.

In L. Koenigsberger
Hermann von Helmholtz
Professor at Heidelberg (pp. 180–1)

Hugo, Victor

Great blunders are often made, like large ropes, or a multitude of fibers.

Les Misérables
Cosette
Book V, Chapter 10

Huxley, Thomas H.

It sounds paradoxical to say the attainment of scientific truth has been effected, to a great extent, by the help of scientific errors.

Method and Results
The Progress of Science (p. 63)

There is no greater mistake than the hasty conclusion that opinions are worthless because they are badly argued.

Method and Result
Natural Rights and Political Rights (p. 369)

. . . irrationally held truths may be more harmful than reasoned errors.

Collected Essays
Volume II
The Coming of Age of "The Origin of Species"

Next to being right in this world, the best of all things is to be clearly and definitely wrong. If you go buzzing about between right and wrong, vibrating and fluctuating, you come out nowhere; but if you are absolutely and thoroughly and persistently wrong, you must, some of these days, have the extreme good fortune of knocking your head against a fact, and that sets you all straight again.

Science and Education
On Science and Art in Relation to Education (p. 174)

Latham, Peter Mere
Amid many possibilities of error, it would be strange indeed to be always in the right.

General Remarks on the Practices of Medicine
The Heart and Its Affections
Chapter IV

It takes as much time and trouble to pull down a falsehood as to build up a truth.

General Remarks on the Practices of Medicine
Chapter VI

It is no easy task to pick one's way from truth to truth through besetting errors.

General Remarks on the Practices of Medicine
Chapter VIII, Part 2

Mach, Ernst
We err when we expect more enlightenment from an hypothesis than from the facts themselves.

The Science of Mechanics (p. 600)

Newton, Sir Isaac
. . . the errors are not the art, but in the artificers.

Mathematical Principles of Natural Philosophy
Preface to the First Edition

Nicolle, Charles
Error is all around us and creeps in at the least opportunity. Every method is imperfect.

In W.I.B. Beveridge
The Art of Scientific Investigation
Difficulties (p. 106)

Steinmetz, Charles Proteus
With the most brilliant engineering design, however, if in the numerical calculation of a single structural member an error has been made and its strength thereby calculated wrong, the rotor of the machine flies to pieces by centrifugal forces, or the bridge collapses, and with it the reputation of the engineer.

Engineering Mathematics (p. 293)

ESTIMATES

King, W.J.
Ideally another man's promises should be negotiable instruments, like his personal check, in compiling estimates.

Mechanical Engineering
The Unwritten Laws of Engineering (p. 4)
June 1944

Much of the difficulty of problem solving comes from the impossibility of getting all the facts together before making a decision.
Edward Hodnett – (See p. 128)

ETHICS

Accreditation Board for Engineering and Technology
Engineers uphold and advance the integrity, honor, and dignity of the engineering profession by:

I   using their knowledge and skill for the enhancement of human welfare;
II being honest and impartial, and serving with fidelity the public, their employer;
III striving to increase the competence and prestige of the engineering profession; and
IV supporting the professional and technical societies of their disciplines.

Code of Ethics for Engineers
October 1977

Florman, Samuel C.
As a professional, I abide by established standards . . . As a human being I hope that I deal adequately with each day's portion of moral dilemmas. But between legality on the one hand and individual predilection on the other, there is hardly any room for the abstraction called 'engineering ethics'.

Blaming Technology
Moral Blueprints (p. 172)

Marshall, T.H.
Ethical codes are based on the belief that between professional and client there is a relationship of trust, and between buyer and seller there is not.

Canadian Journal of Economics and Political Science
The Recent History of Professionalism in Relation
to Social Structure and Social Philosophy (p. 327)
Volume 5, Number 3, August 1939

EXPERIENCE

Adams, Henry Brooks
All experience is an arch to build upon.

The Education of Henry Adams
Rome (p. 87)

Bierce, Ambrose
Experience, *n*. The wisdom that enables us to recognize as an undesirable old acquaintance the folly that we have already embraced.

To One who, journeying through night and fog,
Is mired neck-deep in an unwholesome bog,
Experience, like the rising of the dawn,
Reveals the path that he should not have gone.

Joel Frad Bink
The Enlarged Devil's Dictionary (p. 39)

Bohr, Niels
The great extension of our experience in recent years has brought to light the insufficiency of our simple mechanical conceptions and, as a consequence, has shaken the foundation on which the customary interpretation of observation was based . . .

Atomic Theory and Description of Nature
Introductory Survey (p. 2)

Bowen, Elizabeth
Experience isn't interesting till it begins to repeat itself—in fact, till it does that, it hardly *is* experience.

Death of the Heart
The World (p. 9)

Butler, Samuel
Don't learn to do, but learn in doing. Let your falls not be on a prepared ground, but let them be bona fide falls in the rough and tumble of the world . . .

Samuel Butler's Notebooks (p. 157)

Cardozo, Benjamin N.
Often a liberal antidote of experience supplies a sovereign cure for a paralyzing abstraction built upon a theory.

The Paradoxes of Legal Science
Chapter IV (p. 125)

Cervantes, Miguel de
Experience, the universal Mother of Sciences.

Don Quixote
Part I, Book III, Chapter 7 (p. 140)

Cresy, Edward
The light afforded by Theory will ever appear dim before the Torch of Experience; and well is she represented without hands, and useless, unless guided by practice.

Caption of the 'Address' in
*A Practical Treatise on Bridge Building and on
the Equilibrium of Vaults, and Arches with the
Professional life and Selections from the Works of Rennie*

da Vinci, Leonardo
Nature is full of infinite causes which were never set forth in experience.

The Notebooks of Leonardo da Vinci
Volume I
Philosophy (p. 57)

. . . good judgment is born of clear understanding, and a clear understanding comes of reasons derived from sound rules, and sound rules are the issue of sound experience—the common mother to all the sciences and arts.

The Literary Works of Leonardo da Vinci
Volume I, 18 (p. 119)

But first I shall test by experiment before I proceed further, because my intention is to consult experience first and then with reasoning show why such experience is bound to operate in such a way.

The Literary Works of Leonardo da Vinci
Volume II, 1148 (p. 239)

Experience is never at fault; it is only your judgment that is in error in promising itself such results from experience as are not caused by our experiments. For having given a beginning, what follows from it must necessarily be a natural development of such a beginning, unless it has been subject to a contrary influence, while, if it is affected by any contrary influence, the result which ought to follow from the aforesaid beginning will be found to partake of this contrary influence in a greater or less degree in proportion to the said influence is more or less powerful than the aforesaid beginning.

The Literary Works of Leonardo da Vinci
Volume II, 1153 (p. 240)

Hume, David
. . . all inferences from experience suppose, as their foundation, that the future will resemble the past, and that similar powers will be cojoined with similar sensible qualities. If there be any suspicion that the course of nature may change, and that the past may be no rule for the future, all experience becomes useless, and can give rise to no inference conclusion.

Concerning Human Understanding
Section IV, Part II, 32

Huxley, Aldous
Experience is not a matter of having actually swum the Hellespont, or danced with the dervishes, or slept in a doss-house. It is a matter of sensibility and intuition, of seeing and hearing the significant things, of paying attention at the right moments, of understanding and co-ordinating. Experience is not what happens to a man; it is what a man does with what happens to him.

Texts and Pretexts
Introduction (p. 5)

Latham, Peter Mere
Nothing is so difficult to deal with as man's own Experience, how to value it according to amount, what to conclude from it, and how to use it and do good with it.

In William B. Bean
Aphorisms from Latham (p. 94)

May, Donald C.
In engineering and industry, many important technical decisions are made on the basis of "broad experience" and "sound Judgment". The importance of experience and judgment cannot be overemphasized. However, the technical world is becoming so complicated that these decisions may prove to be expensive and inefficient unless they are backed up by considerable scientific analysis.

Journal of Engineering Education
Choice of Criteria in Applying Mathematical Methods
to Evaluation and Analysis of Engineering Systems (p. 130)
Volume 40, Number 2, October 1949

Roberts, W. Milnor

From the laying out of a line of tunnel to its final completion, the work may be either a series of experiments (made at the expense of the proprietors of the project), or a series of judicious applications of the results of previous experience.

In Henry Drinker
Tunneling, Explosive Compounds and Rock Drills (p. 1005)

Twain, Mark

We should be careful to get out of an experience only the wisdom that is in it—and stop there; lest we be like the cat that sits down on a hot stove-lid. She will never sit down on a hot stove-lid again—and that is well; but she will also never sit down on a cold one anymore.

Following the Equator
Chapter XI
Pudd'nhead Wilson's New Calendar

Unknown

Where did you get your good judgment? Answer: From my experience.

Where did you get your experience? Answer: From my *poor* judgment.

Source unknown

Wilde, Oscar

Experience is the name everyone gives to their mistakes.

Phrases and Philosophies for the Use of the Young
Sebastian Melmoth (p. 85)

EXPERIMENT

Bacon, Francis

The present method of experiment is blind and stupid: hence men wandering and roaming without any determined course, and consulting mere chance, are hurried about to various points, and advance but little . . .

The Novum Organum
First Book, 70

Bernard, Claude

Experiment is fundamentally only induced observation.

An Introduction to the Study of Experimental Medicine
Part I, Chapter I, section v

Beveridge, W.I.B.

. . . no one believes an hypothesis except its originator but everyone believes an experiment except the experimenter.

The Art of Scientific Investigation
Hypothesis (p. 47)

Boyle, Robert

I neither conclude from one single Experiment, nor are the Experiments I make use of, all made upon one Subject: nor wrest I any Experiment to make it quadrare *with* any preconceiv'd Notion. But on the contrary in all kind of Experiments, and all and every one of those Trials, I make the standards (as I may say) or Touchstones by which I try all my former Notions, whether they hold not in weight or measure and touch, Ec. For as that Body is no other than a Counterfeit Gold, which wants any one of the Properties of Gold (such as are the Malleableness, Weight, Colour, Fixtness in the Fire, Indissolubleness in Aquafortis, and the like), though it has all the other; so will all those notions be found false and deceitful, that will not undergo all the Trials and Tests made of them by Experiments. And therefore such as will not come up to the desired Apex of Perfection, I rather wholly reject and take a new, than by piercing and patching endeavour to retain the old, as knowing old things to be rather made worse by mending the better.

In Michael Roberts and E.R. Thomas
Newton and the Origin of Colours (p. 53)

da Vinci, Leonardo
But first I will make some experiment before proceeding farther because it is my intention first to cite experience then to show by reasoning why this experience is constrained to act in this manner. And this is the rule according to which speculators as to natural effects have to proceed. And although nature commences with reason and ends in experience it is necessary for us to do the opposite, that is to commence as I said before with experience and from this proceed to investigate the reason.

The Notebooks of Leonardo da Vinci
Volume I, Chapter XIX

de Fontenelle, Bernard
'Tis no easy matter to be able to make an Experiment with accuracy. The least fact, which offers itself to our consideration, takes in so many other facts, which modify or compose it, that it requires the utmost dexterity to lay open the several branches of its composition, and no less sagacity to find 'em out.

In Michael Roberts and E.R. Thomas
Newton and the Origin of Colours (p. 6)

Dubos, René
The experiment serves two purposes, often independent one from the other: it allows the observation of new facts, hitherto either unsuspected, or not yet well defined; and it determines whether a working hypothesis fits the world of observable facts.

In W.I.B. Beveridge
The Art of Scientific Investigation
Experimentation (p. 13)

Eddington, Sir Arthur Stanley
. . . he is an incorruptible watch-dog who will not allow anything to pass which is not observationally true.

The Philosophy of Physical Science (p. 112)

Edison, Thomas
The only way to keep ahead of the procession is to experiment. If you don't, the other fellow will. When there's no experimenting there's no progress. Stop experimenting and you go backward. If anything goes wrong, experiment until you get to the very bottom of the trouble.

In Frank Lewis Dyer
Edison: His Life and Inventions
Volume II, Chapter XXIV (p. 617)

Ehrlich, Paul
Much testing; accuracy and precision in experiment; no guesswork or self-deception.

In Martha Marquardt
Paul Ehrlich
Chapter XIII (p. 134)

Gregg, Alan
Experiments are like cross-questioning a witness who will tell the truth but not the whole truth.

The Furtherance of Medical Research
Chapter III (p. 89)

Maxwell, James Clerk
An Experiment, like every other event which takes place, is a natural phenomenon; but in a Scientific Experiment the circumstances are so arranged that the relations between a particular set of phenomena may be studied to the best advantage. In designing an Experiment the agents and the phenomena to be studied are marked off from all others and regarded as the Field of Investigation.

The Scientific Papers of James Clerk Maxwell (p. 505)

Planck, Max
Experimenters are the shocktroops of science.

Scientific Autobiography (p. 110)

An experiment is a question which science poses to Nature, and a measurement is the recording of Nature's answer.

Scientific Autobiography (p. 110)

Poincaré, Henri
Experiment is the sole source of truth. It alone can teach us anything new; it alone can give us certainty.

Science and Hypothesis
Chapter IX

Robertson, Howard P.
What is needed is a homely experiment which could be carried out in the basement with parts from an old sewing machine and an Ingersol watch, with an old file of *Popular Mechanics* standing by for reference!

In Paul Arthur Schlipp
Albert Einstein: Philosopher–Scientist
Geometry as a Branch of Physics (p. 326)

Thomson, Sir George
. . . in order to make an experiment meaningful one must have a theory as to what matters for the experiment.

The Inspiration of Science
Chapter II (p. 15)

Truesdell, Clifford A.
The hard facts of classical mechanics taught to undergraduates today are, in their present forms, creations of James and John Bernoulli, Euler, Lagrange, and Cauchy, men who never touched a piece of apparatus; their only researches that have been discarded and forgotten are those where they tried to fit theory to experimental data. They did not disregard experiment; the parts of their work that are immortal lie in domains where experience, experimental or more common, was at hand, already partly understood through various special theories, and they abstracted and organized it and them. To warn scientists today not to disregard experiment is like preaching against atheism in church or communism among congressmen. It is cheap rabble-rousing. The danger is all the other way. Such a mass of experimental data on everything pours out of organized research that the young theorist needs some insulation against its disrupting, disorganizing effect. Poincaré said, "The scientist must order; science is made out of facts as a house is made out of stones, but an accumulation of facts is no more science than a heap of stones, a house." Today the houses are buried under an avalanche of rock splinters, and what is called theory is often no more than the trace of some moving fissure on the engulfing wave of rubble. Even in earlier times there are examples. Stokes derived from his theory of fluid friction the formula for the discharge from a circular pipe. Today this classic formula is called the "Hagen–Poiseuille law" because Stokes, after comparing it with measured data and finding it did not fit, withheld publication. The data he had seem to have concerned turbulent flow, and while some experiments that confirm his mathematical discovery had been performed, he did not know of them.

Six Lectures on Modern Natural Philosophy
Method and Taste in Natural Philosophy (pp. 92–3)

Weyl, Hermann
Allow me to express now, once and for all, my deep respect for the work of the experimenter and for his fight to wring significant facts from an inflexible Nature, who says so distinctly "No" and so indistinctly "Yes" to our theories.

The Theory of Groups and Quantum Mechanics (p. xx)

FACTS

Abbott, Edwin A.
From dreams I proceed to facts.

Flatland (p. 68)

Adams, Henry Brooks
The facts seemed certain, or at least as certain as other facts; all they needed was explanation.

The Education of Henry Adams
The Abyss of Ignorance (p. 435)

Alcott, Louisa May
Entrenching himself behind an undeniable fact.

Little Women
XXXV

Aristotle
. . . with a true view all the data harmonize, but with a false one the facts soon clash.

Nicomachean Ethics
Book I, Chapter VIII

Balchin, Nigel
"Am I supposed to give all the facts, or some of the facts, or my opinions or your opinions or what?"

The Small Back Room (p. 53)

Barry, Frederick
A fact is no simple thing.

The Scientific Habit of Thought (p. 92)

Baruch, Bernard M.
If you get all the facts, your judgment can be right; if you don't get all the facts, it can't be right.

St. Louis Post Dispatch
21 June 1965 (p. 5a)

Cross, Hardy
In so far as engineers deal with facts that can be measured they use mathematics to combine these facts and to deduce their conclusions. But often the facts are not subject to exact measurement or else the combinations are of facts that are incommensurable.

Engineers and Ivory Towers
The Education of an Engineer (p. 64)

The constant and insistent need that engineers feel for any scrap of fact from which they may predict natural phenomena tends to develop a hunger for anything that even resembles a fact. This in turn may lead to a wolfish and gluttonous attitude, a gobbling up of every statement or opinion, figure or formula, indiscriminately and incessantly. The result is often intellectual autointoxication from "hunks and gobs" of unselected, undigested and indigestible material.

Engineering and Ivory Towers
For Man's Use of God's Gifts (p. 99)

. . . engineers need to select their mental diet carefully and when they go a-fishing after facts they want a fish fry and not a chowder.

Engineering and Ivory Towers
For Man's Use of God's Gifts (p. 99)

Einstein, Albert
The justification for a physical concept lies exclusively in its clear and unambiguous relation to facts that can be experienced.

In A.P. French
Einstein: A Centenary Volume (p. 229)

Faraday, Michael
I could trust a fact and always cross-question an assertion.

In Oswald Blackwood
Introductory College Physics (p. 413)

Gardner, Earl Stanley
Facts themselves are meaningless. It's only the interpretation we give those facts which counts.

The Case of the Perjured Parrot (p. 171)

Greenstein, Jesse L.
Knowing how hard it is to collect a fact, you understand why most people want to have some fun analyzing it.

Fortune
Great American Scientists: The Astronomer (p. 149)
Volume 61, Number 5, May 1960

Hodnett, Edward
Much of the difficulty of problem solving comes from the impossibility of getting all the facts together before making a decision.

The Art of Problem Solving (p. 43)

A fact not recognized for what it signifies has no more value than a precious stone in a savage's collection of shells and pebbles.

The Art of Problem Solving (p. 43)

Ignorance of the significance of facts renders us as blind to the solution of a problem as if we were matching colors in the dark.

The Art of Problem Solving (p. 43)

The necessity for getting facts straight leads the professional problem solver to take what seems to the layman fantastic pains in checking even small details.

The Art of Problem Solving (p. 48)

What you call a fact may with good reason not seem a fact to the other fellow.

The Art of Problem Solving (p. 50)

Huxley, Aldous
Facts do not cease to exist because they are ignored.

Proper Studies
A Note on Dogma (p. 205)

Huxley, Julian
To speculate without facts is to attempt to enter a house of which one has not the key, by wandering aimlessly round and round, searching the walls and now and then peeping through the windows. Facts are the key.

Essays in Popular Science
Heredity
I, The Behavior of the Chromosomes (pp. 1–2)

Huxley, Thomas H.
Sit down before fact as a little child, follow humbly wherever and whatever abysses nature leads, or you shall learn nothing.

The Life and Letters of Thomas Henry Huxley
Volume I
Letter written on 23 September 1860 (p. 219)

Those who refuse to go beyond fact rarely get as far as fact.

Methods and Results
The Process of Science (p. 62)

Latham, Peter Mere
Bear in mind then, that abstractions are *not facts*; and next bear in mind that *opinions* are not facts.

In William B. Bean
Aphorisms from Latham (p. 36)

McCarthy, Mary
. . . in science, all facts, no matter how trivial or banal, enjoy democratic equality.

On the Contrary
The Fact in Fiction (p. 266)

Oppenheimer, Julius Robert
. . . when technical people talk they always emphasize the facts that they are not sure.

Harper's Magazine
The Tree of Knowledge (p. 57)
Volume 217, October 1958

Polanyi, Michael
Just as the eye sees details that are not there if they fit in with the sense of the picture, or overlooks them if they make no sense, so also very little inherent certainty will suffice to secure the highest scientific value to an alleged fact, if only it fits in with a great scientific generalization, while the most stubborn facts will be set aside if there is no place for them in the established framework of science.

Personal Knowledge
Chapter 6 (p. 138)

Siegel, Eli
Facts are always whispering, uttering, and shouting advice.

Damned Welcome (p. 152)

The facts never give up.

Damned Welcome (p. 156)

Smith, George Otis

. . . statistics need to be much more than the output of a battery of adding machines. The ideal collection of facts is the man who has spent years as a specialist in the work and in this way knows the reality behind the words and figures. Only the personal touch that comes from intimate familiarity with facts at their source can give life to statistics.

Civil Engineering
What are the Facts? (p. 154)
Volume 2, Number 3, March 1932

Facts that have aged in the course of their collection and preparation for consumption are likely to be too stale for practical use. Dating an egg doesn't keep it good.

Civil Engineering
What are the Facts? (p. 155)
Volume 2, Number 3, March 1932

In this partnership of engineer and economist, it will be the engineer's part to furnish most of the facts. The engineer calls them "plain" facts, because they do not lend themselves to display as readily as theoretical phrases. He uses facts, not pieces on a chessboard to be moved back and forth in a contest of wits, but rather as foundation stones to be assembled in orderly fashion to hold up the superstructure of conclusions.

Civil Engineering
What are the Facts? (p. 155)
Volume 2, Number 3, March 1932

Twain, Mark

Get your facts first, then you can distort them as much as you please.

In Rudyard Kipling
From Sea to Sea
An Interview with Mark Twain

Unknown

My mind is like a coal chute down which many tons of facts have rumbled, leaving only a little dust behind.

In Edward Hodnett
The Art of Problem Solving (p. 42)

FAILURE

Bayard, Herbert
I cannot give you the formula for success, but I can give you the formula for failure—which is: Try to please everyone.

Address
Interfaith in Action
20 December 1950

Huxley, Thomas H.
. . . there is the greatest practical benefit in making a few failures early in life.

Science and Education
On Medical Education (p. 306)

Ruskin, John
Failure is less attributable to either insufficiency of means or impatience of labours than to a confused understanding of the thing actually to be done.

In Henry Attwell
Thoughts from Ruskin (p. 12)

FLUID

Lamb, Sir Horace

I am an old man now, and when I die and go to Heaven there are two matters on which I hope for enlightenment. One is quantum electrodynamics and the other is the turbulent motion of fluids. And about the former I am rather more optimistic.

Attributed
In Dale A. Anderson, John C. Tennehill and Richard H. Pletcher
Computational Fluid Mechanics and Heat Transfer (p. 197)

FORCE

Bethe, Hans
The concept of force will continue to be effective and useful.

Source unknown

Hembree, Lawrence
Two first graders were standing outside the school one morning.

"Do you think," said one, "that thermo-nuclear projectiles will be affected by radiation belts?"

"No," replied the other, "Once a force enters space . . ."

The school bell rang, "Damn it," said the first kid. "Here we go, back to the old bead stringing."

Quote
Light Armor (p. 118)
Volume 54, Number 6, 6 April 1967

Maxwell, James Clerk
An inextensible heavy chain
Lies on a smooth horizontal plane,
An impulsive force is applied at A,
Required the initial motion of K.

In Lewis Campbell and William Garnett
The Life of James Clerk Maxwell
A Problem in Dynamics (p. 625)

Gin a body meet a body
   Flyin' through the air,
Gin a body hit a body,
   Will it fly? and where?

In Lewis Campbell and William Garnett
The Life of James Clerk Maxwell
In Memory of Edward Wilson (p. 630)

FORECAST

Fiedler, Edgar R.
If you have to forecast, forecast often.

Across the Board
The Three Rs of Economic Forecasting—Irrational, Irrelevant and Irreverent
June 1977

The herd instinct among forecasters makes sheep look like independent thinkers.

Across the Board
The Three Rs of Economic Forecasting—Irrational, Irrelevant and Irreverent
June 1977

He who lives by the crystal ball soon learns to eat ground glass.

Across the Board
The Three Rs of Economic Forecasting—Irrational, Irrelevant and Irreverent
June 1977

The moment you forecast you know you're going to be wrong, you just don't know when and in which direction.

Across the Board
The Three Rs of Economic Forecasting—Irrational, Irrelevant and Irreverent
June 1977

Harris, Ralph
All forecasting is in an important sense *backward*-looking—vividly compared to steering a ship by its wake.

Economic Forecasting—Models or Markets? (p. 86)

Penjer, Michael
We are making forecasts with bad numbers, but the bad numbers are all we've got.

The New York Times
1 September 1989

FORMULA

Barr, H.F.
The young engineer, for example, soon finds that a problem is not always clear or easily defined and that the solution does not involve merely substituting known values into a standard formula.

General Motors Engineering Journal
Typical Problems in Engineering
Foreword
Set 1, Number 1

Dudley, Underwood
Formulas should be useful. If not they should be astounding, elegant, enlightening, simple, or have some other redeeming value.

Mathematical Magazine
Formulas for Primes
Volume 56, 1983

Hodnett, Edward
A formula is like a basket. Try to pick up a dozen apples from the ground and carry them in your hands. It is well-nigh impossible. With a basket you can carry as many as you can lift.

The Art of Problem Solving (p. 86)

What you do to a situation when you use a formula approach is to schematize it. You impose a pattern on it . . .

The Art of Problem Solving (p. 89)

Peirce, Charles Sanders
It is terrible to see how a single unclear idea, a single formula without meaning, lurking in a young man's head, will sometimes act like an obstruction of inert matter in an artery, hindering the nutrition of the brain, and condemning its victim to pine away in the fullness of his intellectual vigor and in the midst of intellectual plenty.

Chance, Love and Logic
Part I, Second Paper (p. 37)

Saint Augustine

If I am given a formula, and I am ignorant of its meaning, it cannot teach
me anything, but if I already know it what does the formula teach me?

De Magistro
Chapter X, 23

Woll, Matthew

I know very well that in a great many circles the man who does not enter
with a neatly arranged plan, with a set of doctrines, with a rounded and
sonorous formula, and with assurance about everything, is set down as
something of an old fogey, perhaps reactionary, certainly not one of the
elect who are "doing things" and providing guidance for the race. I must
assume the risk. I have no formula. [But I shall resist] those who have
the formula for so many things and who seek so avidly to force it down
the throats of every one else.

Annals (American Academy of Political and Social Science)
Standardization (p. 47)
Volume 137, May 1928

A formula is like a basket...
Edward Hodnett – (See p. 135)

FRICTION

Unknown
Frictional effects are a real drag.

Source unknown

GENIUS

Amiel, Henri Frédéric
To do easily what is difficult for others is the mark of talent. To do what is impossible for talent is the mark of genius.

Journal Intime (p. 76)

Austin, Henry
Genius, that power which dazzles mortal eyes,
Is oft but perseverance in disguise.

Perseverance Conquers All

Bartol, C.A.
As diamond cuts diamond, and one hone smooths a second, all the parts of intellect are whetstones to each other; and genius which is but the result of their mutual sharpening is character too.

Radical Problems
Individualism (p. 43)

Edison, Thomas
Genius is 1 per cent inspiration and 99 per cent perspiration.

In Frank Lewis Dyer
Edison: His Life and Inventions
Volume II, Chapter XXIV (p. 607)

Fitzgerald, F. Scott
Genius is the ability to put into effect what is in your mind. There's no other definition of it.

The Crack-Up
The Note-Books
E (p. 123)

Whipple, E.P.

Talent repeats; Genius creates. Talent is a cistern; Genius is a fountain. Talent deals with the actual, with discovered and realized truths, analyzing, arranging, combining, applying positive knowledge, and in action looking to precedents; Genius deals with the possible, creates new combinations, discovers new laws, and acts from an insight into principles. Talent jogs to conclusions to which Genius takes giant leaps. Talent accumulates knowledge, and has it packed up in the memory; Genius assimilates it with its own substance, grows with every new accession, and converts knowledge into *power*. Talent gives out what it has taken in; Genius what has risen from its unsounded wells of living thought. Talent, in difficult situations, strives to untie knots, which Genius instantly cuts with one swift decision. Talent is full of thoughts, Genius of thought; one has definite acquisitions, the other indefinite power.

Literature and Life
Genius (p. 162)

Genius is 1 per cent inspiration and 99 per cent perspiration.
Thomas Edison – (See p. 138)

GEOLOGY

Dawkins, Boyd
Geology stands to engineering in the same relation as faith to works . . .
The success or failure of an undertaking depends largely upon the
physical conditions which fall within the province of geology, and the
"works" of the engineer should be based on the "faith" of the geologist.

Inst. Civil Eng. Min. Proc. (London)
On the Relation of Geology to Civil Engineering (pp. 254–255)
Volume 134, 1898

Trefethen, Joseph M.
The relationship between civil engineering and geology is as old as the
hills, manmade hills that is.

Journal of Engineering Education
Geology for Civil Engineers (p. 383)
Volume 39, Number 7, March 1949

GOAL

Farber, Eric A.
Everything worthwhile which has been achieved by an individual, a group, or humanity as a whole was done by striving toward a definite, clearly defined—although sometimes unattainable—goal.

Journal of Engineering Education
The Teaching and Learning of Engineering (p. 784)
Volume 45, Number 10, June 1955

Fredrickson, A.G.
We must try to set up some definite goals that have the benefit of all mankind as their objective.

Chemical Engineering Education
The Dilemma of Innovating Societies (p. 148)
Volume 4, Summer 1969

Only by setting up defined goals will it be possible to develop priorities and institutions that can guide the innovative genius of men onto paths that will be truly, as opposed to superficially, beneficial.

Chemical Engineering Education
The Dilemma of Innovating Societies (p. 148)
Volume 4, Summer 1969

GRAPHICS

Anthony, Gardner C.
Any collection of related facts is difficult to grasp when expressed by figures in tabular form, but the same may be seen at a glance when presented by one of the many graphic representations of those ideas.

An Introduction to the Graphic Language (p. iii)

Rogers, Will
You must never tell a thing. You must illustrate it. We learn through the eye and not the noggin.

The Will Rogers Book
25 June 1933 (p. 121)

GRAVITY

Bronowski, Jacob
There are two experiences on which our visual world is based: that gravity is vertical, and that the horizon stands at right angles to it.

The Ascent of Man (p. 157)

Emerson, Ralph Waldo
The child amidst his baubles is learning the action of light, motion, gravity . . .

The Works of Ralph Waldo Emerson
Volume I
Address (p. 121)

Unknown
What goes up must come down.

Source unknown

HEAT

Flanders, Michael
Swann, Donald
You can't pass heat from a cooler to a hotter.
Try if you like, you far better notter,
'cause the cold in the cooler will get hotter as a ruler,
'cause the hotter body's heat will pass to the cooler.

At the Drop of Another Hat
The First and Second Law

Keane, Bill
Heat makes things expand. That's why the days are longer in the summer.

Caption to cartoon

McNeil, I.
"In the beginning God created the heaven and the earth" are the opening words of the Bible, which goes on, in verse 3, "and God said Let there be light: and there was light". It must be assumed that the Cosmic Illuminator had to abide by the laws of physics like the rest of us for, after all, He had created them. In the nature of things, sensible heat comes long before visible light in the spectrum of electromagnetic wavelengths. Thus, when God said "Let there be light", he implied, "Let there also be heat"—and there was heat.

In R. Angus Buchanan
Engineers and Engineering
Blast: From Blowpipe to Blowing Engine (p. 79)

Mayer, Julius Robert von
Concerning the intimate nature of heat, or of electricity, etc, I know nothing, any more than I know the *intimate nature* of any matter whatsoever, or of anything else.

Kleinere Schriften und Briefe (p. 181)
In Pierre Duhem
The Aim and Structure of Physical Theory (p. 52)

144

Metsler, William
The stove is hot, but that's no change
Heat's what it's supposed to make
Resistance generates the energy to bake
So its always Ohm Ohm on the range.

The Physics Teacher
The Cowboy's Lament (p. 127)
Volume 15, Number 2, February 1977

Mott-Smith, Morton
People are always exaggerating temperatures. If the day is hot, they add on a few degrees; if it is cold they deduct a few. No one ever gives the air temperature to a fraction of a degree, but only to whole degrees. Now on the Fahrenheit scale, on account of the small size of its degree, these whoppers and inaccuracies are only about half as big as they are on the other scales.

Heat and Its Workings (p. 24)

Poincaré, Henri
And a well-made language is no indifferent thing; not to go beyond physics, the unknown man who invented the word *heat* devoted many generations to error. Heat has been treated as a substance, simply because it was designated by a substance, and it has been thought indestructible.

The Foundations of Science
The Value of Science
Analysis and Physics (p. 289)

IDEA

Bagehot, Walter

One of the greatest pains to human nature is the pain of a new idea.

Physics and Politics (p. 163)

Bernard, Claude

If an idea presents itself to us, we must not reject it simply because it does not agree with the logical deductions of a reigning theory.

An Introduction to the Study of Experimental Medicine
Part 1, Chapter 2, section iii

Our ideas are only intellectual instruments which we use to break into phenomena; we must change them when they have served their purpose, as we change a blunt lancet that we have used long enough.

An Introduction to the Study of Experimental Medicine
Part I, Chapter 2, section iv

Butler, Samuel

Every idea has something of the pain and peril of childbirth about it; ideas are just as mortal and just as immortal as organized beings are.

Samuel Butler's Notebooks
c

Dobshansky, Theodosius

Great ideas often seem simple and self-evident, but only after somebody has explained them to us. Then, how interesting they become! The act of insight is among the most exciting and pleasurable experiences a scientist can have, when he recognizes what all the time was there to be seen, and yet he did not see it.

In Robert M. Hutchins and Mortimer J. Adler (Editors)
The Great Ideas
1974
Advancement and Obsolescence in Science (p. 56)

146

Easton, William
A creative thinker evolves no new ideas. He actually evolves new combinations of ideas that are already in his mind.

In Alex F. Osborn
Applied Imagination (p. 173)

Eliot, George
The moment of finding a fellow-creature is often as full of mingled doubt and exultation as the moment of finding an idea.

Daniel Deronda
Chapter 17

Eliot, T.S.
Between the idea
And the reality
Between the motion
And the act
Falls the Shadow.

Collected Poems 1909–1935
The Hollow Men
V

Emerson, Ralph Waldo
God screens us evermore from premature ideas.

Essays
Spiritual Laws

Hoefler, Don C.
Develop a honeybee mind, gathering ideas everywhere and associating them fully.

Electronic News
But You Don't Understand the Problem
17 July 1967

Holmes, Sherlock
Our ideas must be as broad as Nature if they are to interpret Nature.

In Arthur Conan Doyle
The Complete Sherlock Holmes
A Study in Scarlet, Chapter 5

Huxley, Thomas H.
. . . whatever practical people may say, this world is, after all, absolutely governed by ideas, and very often by the wildest and most hypothetical ideas.

Science and Education
On the Study of Biology (p. 273)

James, William
An idea, to be suggestive, must come to the individual with the force of a revelation.

Varieties of Religious Experience (p. 113)

Kettering, Charles Franklin
I have no objection to the standardizing of bolts and nuts and screws . . . but I do have a terrible obsession against the standardization of ideas.

In T.A. Boyd
Professional Amateur (p. 137)

Locke, John
It is not in the power of the most exhalted wit or enlarged understanding, by any quickness or variety of thought, to invent or frame one new simple idea.

Essay Concerning Human Understanding
II

There seems to be a constant decay of all our ideas; even of those which are struck deepest and in minds the most retentive, so that if they be not sometimes renewed by repeated exercises of the senses, or reflection on those kinds of objects which at first occasioned them, the print wears out, and at last there remains nothing to be seen.

Essay Concerning Human Understanding
II

Milne, A.A.
When you are a Bear of Very Little Brain, and you Think of Things, you find sometimes that a Thing which seemed very Thingish inside you is quite different when it gets out into the open and has other people looking at it.

The House at Pooh Corner

Rossman, Joseph
One seldom perfects an idea without many failures . . .

Industrial Creativity: The Psychology of the Inventor (p. 45)

Shockley, William B.
If you have a bright idea and you do the right kind of experiment, you may get pretty decisive results pretty soon.

Esquire
1973

Siegel, Eli
An idea is an eddy, an island of the mind, connected with a vast mainland.

Damned Welcome (p. 157)

Spencer, Herbert
Early ideas are not usually true ideas.

Principles of Biology
Part ii, Chapter 2, section 110

Thomson, James
Ten thousand great ideas filled his mind;
But with the clouds they fled, and left no trace behind.

Castle of Indolence
Canto 1, stanza 59

Trotter, Wilfred
The mind likes a strange idea as little as the body likes a strange protein, and resists it with similar energy. It would not be too fanciful to say that a new idea is the most quickly acting antigen known to science. If we watch ourselves honestly we shall often find that we have begun to argue against a new idea before it has been completely stated.

Collected Papers
Has the Intellect a Function?

Wells, H.G.
He had ideas about everything. He could no more help having ideas about everything than a dog can resist smelling at your heels.

Mr. Britling Sees It Through
Book 1, section 2

IMAGINATION

Bradley, A.C.
Research, though toilsome, is easy; imaginative vision, though delightful, is difficult.

Oxford Lectures on Poetry
Shakespeare's Theatre and Audience

Beveridge, W.I.B.
The imagination merely enables us to wander into the darkness of the unknown where, by the dim light of the knowledge that we carry, we may glimpse something that seems of interest. But when we bring it out and examine it more closely it usually proves to be only trash whose glitter had caught our attention. Imagination is at once the source of all hope and inspiration but also of frustration. To forget this is to court despair.

The Art of Scientific Investigation (p. 5)

Dewey, John
Every great advance in science has issued from a new audacity of imagination.

The Quest for Certainty
Chapter XI (p. 310)

IMPOSSIBLE

Aristotle
What is convincing though impossible should always be preferred to what is possible and unconvincing.

The Poets
Chapter XXIV

Carroll, Lewis
Alice laughed. "There's no use trying," she said "one can't believe impossible things."

"I daresay you haven't had much practice," said the Queen. "When I was your age, I always did it for half-an-hour a day. Why, sometimes I've believed as many as six impossible things before breakfast."

The Complete Works of Lewis Carroll
Through the Looking Glass
Wool and Water

Clarke, Arthur C.
The only way of finding the limits of the possible is by going beyond them into the impossible.

The Lost Worlds of 2001
Chapter 34 (p. 189, fn)

Goddard, Robert H.
It is difficult to say what is impossible, for the dream of yesterday is the hope of today and the reality of tomorrow.

Technology and Culture
In Eugene M. Emme
Introduction to the History of Rocket Technology (p. 377)
Fall 1963

Juster, Norton

. . . so many things are possible just as long as you don't know they're impossible.

The Phantom Tollbooth (p. 247)

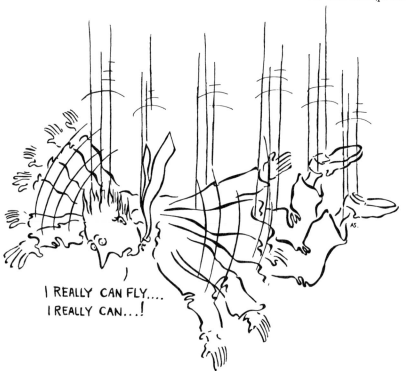

I REALLY CAN FLY....
I REALLY CAN...!

Wilde, Oscar

Man can believe the impossible, but man can never believe the improbable.

Phrases and Philosophies for the Use of the Young
Sebastian Melmoth (p. 123)

IMPRESSION

Bacon, Francis
The human understanding is most excited by that which strikes and enters the mind at once and suddenly, and by which the imagination is immediately filled and inflated. It then begins almost imperceptibly to conceive and suppose that everything is similar to the few objects which have taken impression on the mind.

In W.I.B. Beveridge
The Art of Scientific Investigation
Difficulties (p. 118)

INFORMATION

Franklin, Benjamin

I find a frank acknowledgment of one's ignorance is not only the easiest way to get rid of a difficulty, but the likeliest way to obtain information, and therefore I practice it: I think it an honest policy. Those who affect to be thought to know everything, and so undertake to explain everything, often remain long ignorant of many things that others could and would instruct them in, if they appeared less conceited.

The Works of Benjamin Franklin
Volume I
Electricity (p. 307)

Unknown

1. The information you have is not what you want.
2. The information you want is not what you need.
3. The information you need is not what you can obtain.
4. The information you can obtain costs more than you want to pay!

Laws of Information
Source unknown

INNOVATION

Fredrickson, A.G.

The point is simply that if a technological innovation has a good side (as it almost always does), it will more than likely have a bad side as well.

Chemical Engineering Education
The Dilemma of Innovating Societies (p. 144)
Summer 1969

Gardner, John W.

We may learn something about the renewal of societies if we look at the kind of men who contribute most to that outcome—the innovators.

Self-renewal: The Individual and the Innovative Society (p. 27)

Many of the major changes in history have come about through successive small innovations, most of them anonymous.

Self-renewal: The Individual and the Innovative Society (p. 31)

INSPIRATION

Hugo, Victor

One must fill oneself with human science. Above all and in spite of all, be a man. Do not fear to surcharge yourself with humanity. Ballast your mind with reality and then throw yourself into the sea—the sea of inspiration.

Intellectual Autobiography (pp. 124–5)

Poincaré, Henri

One need only open the eyes to see that the conquest of industry which has enriched so many practical men would never have seen the light, if these practical men alone had existed and if they had not been preceded by unselfish devotees who died poor, who never thought of utility, and yet had a guide far other than caprice.

The Foundations of Science
Science and Method
Science and the Scientist (p. 363)

INVENTIONS

Bacon, Francis
The human mind is often so awkward and ill-regulated in the career of invention that is at first diffident, and then despises itself. For it appears at first incredible that any such discovery should be made, and when it has been made, it appears incredible that it should so long have escaped men's research. All which affords good reason for the hope that a vast mass of inventions yet remains, which may be deduced not only from the investigation of new modes of operation, but also from transferring, comparing and applying those already known, by the methods of what we have termed literate experience.

Novum Organum
Book I

Bierce, Ambrose
An inventor is a person who makes an ingenious arrangement of wheels, levers and springs, and believes it civilization.

The Enlarged Devil's Dictionary

Brecht, Bertolt
The more we can squeeze out of nature by inventions and discoveries and improved organization of labour, the more uncertain our existence seems to be. It's not we who lord it over things, it seems, but things which lord it over us.

The Messingkauf Dialogues
The Second Night
The Philosopher's Speech about our Period (p. 42)

Disraeli, Isaac
The golden hour of invention must terminate like other hours, and when the man of genius returns to the cares, the duties, the vexations, and the amusements of life, his companions behold him as one of themselves—the creature of habits and infirmities.

Literary Characters of Men of Genius
Chapter XVI

Drachmann, A.G.
I should prefer not to seek for the cause of the failure of an invention in the social conditions till I was quite sure that it was to be found in the technical possibilities of the time.

The Mechanical Technology of Greek and Roman Antiquity (p. 206)

Dyer, Frank Lewis
Martin, Thomas Commerford
All persons who make inventions will necessarily be more or less original in character, but to the man who chooses to become an inventor by profession must be conceded a mind more than ordinarily replete with virility and originality.

Edison: His Life and Inventions
Volume II, Chapter XXIV (p. 597)

Emerson, Ralph Waldo
Invention breeds invention.

Society and Solitude
Works and Days

In America, the geography is sublime, but the men are not; the inventions are excellent, but the inventors one is sometimes ashamed of.

The Works of Ralph Waldo Emerson
Volume VI
Conduct of Life
Considerations by the way

Galbraith, John Kenneth
Inventions that are not made, like babies that are not born, are rarely missed.

The Affluent Society
Chapter 9, section III (p. 105)

Hamilton, Walton
Till, Irene
Most discoveries patented today can be anticipated . . . For the most part, technicians are not self starters. The bulk of them in fact are captives; those in corporate employ are told by business executives what problems to work on . . . The solo inventor's real opportunity is to seize or blunder upon a pioneer idea; as a technology foliates from its base, his self reliance is hardly a match for a bevy of experts who can be kept on the job . . . A captive technology offers no chance to invent except to those already in control, or to others on such terms as those in control dictate.

Law and Contemporary Problems
Volume 13, 1948 (p. 252)

Hubbard, Elbert
The brains of a thousand inventors have seethed, dreamed, contrived, thought, so as to bring me to my present form.

Notebook (p. 116)

In an inventor's work there is required something similar to that which the artist brings to bear.

Notebook (p. 194)

Huxley, Aldous
NINETEENTH-CENTURY science discovered the technique of discovery, and our age is, in consequence, the age of inventions. Yes, the age of inventions: we are never tired of proclaiming the fact. The age of inventions—and yet nobody has succeeded in inventing a new pleasure.

Music at Night and Other Essays (p. 136)

Kaempffert, W.B.
It is not difficult to predict the effect of industrial group research on invention. As organized invention and discovery gain momentum the revolutionist will have no chance in explored fields. He will have to compete with more and more men who have at their disposal splendidly equipped laboratories, time and money, and who may work for three or four years before producing a noteworthy result. Only the exceptionally brilliant trained scientist will be able to meet these explorers on their own ground. Possibly Edison may be the last of the great heroes of invention.

Invention and Society (p. 30)

Kettering, Charles Franklin
The lack of ideas and inventions in one generation can easily mean the loss of Freedom in the next.

Short Stories of Science and Invention (p. 35)

Keyser, Cassius J.
But for the unattainable ideal of perfect conquest of Nature, we should not have the marvels of modern Invention and Empirical Science.

Mole Philosophy & Other Essays (p. 3)

King, Blake
. . . if your conclusion is that the operator should be able to make daily adjustments on your invention by hitting it with a rock, say so.

Mechanical Engineering
Object: Creativity (p. 41)
November 1963

Lemelson, Jerome
An important part of invention today is being able to discover the problem.

In Kenneth A. Brown
Inventors at Work (p. 126)

Maclaurin, W.R.
We have now reached a stage in many fields where inventions are almost made to order, and where there can be a definite correlation between the number of applied scientists employed (and the funds at their disposal) and the inventive results. But one really gifted inventor is likely to be more productive than half a dozen men of lesser stature.

Quarterly Journal of Economics
The Sequence from Invention to Innovation
February 1953

Marcellus
The principle reasons which have been assigned for the slow progress of the useful arts, are drawn from the wide separation which has been made between science and art, and the fact that many of the greatest inventions have resulted from accident rather than superior knowledge. These considerations have tended to give to these pursuits the character of unintellectual employments.

Young Mechanic
Essay on the Mechanic Arts (p. 36)
Volume I, Number III, March 1832

Marconi, G.
Necessity is the cause of many inventions but the best ones are born of desire.

Colliers
Every Man His Own Inventor
Volume 70, 1922

Marshall, Alfred
The full importance of an epoch-making idea is often not perceived in the generation in which it is made . . . The mechanical inventions of every age are apt to be underrated relatively to those of earlier times. For a new discovery is seldom fully effective for practical purposes till many minor improvements and subsidiary discoveries have gathered themselves around it.

Principles of Economics
8th Edition
Book IV, Chapter VI, I (p. 205, fn)

McArthur, Peter
It was a mere detail that my invention was no good.

The Best of Peter McArthur
The Great Experiment (p. 26)

Middendorf, W.H.
Brown, G.T., Jr.
The romantic theory that an invention will appear in full bloom without conscious effort on the part of the gifted inventor has been deprecated.

Electrical Engineering
Orderly Creative Inventing (p. 861)
October 1957

Mumford, Lewis
By his very success in inventing labor-saving devices, modern man has manufactured an abyss of boredom that only the privileged classes in earlier civilizations have ever fathomed . . .

Conduct of Life
The Challenge of Renewal (p. 14)

Oppenheimer, Julius Robert
Manifestly not every finding leads straight to invention; but it is hard to think of major discoveries about nature; major advances in science, which have not had large and ramified practical consequences.

In Dael Wolfle (Editor)
Symposium on Basic Research
The Need for New Knowledge (p. 6)

Philip, Prince, Duke of Edinburgh
Innovation depends on invention and inventors should be treated as the pop stars of industry.

Attributed

Psalms 106:29
Thus they provoked *him* to anger with their inventions: and the plague brake in upon them.

The Bible

Psalms 106:39
Thus they were defiled with their own works, and went a whoring with their own inventions.

The Bible

Swift, Jonathan
The greatest Inventions were produced in the Times of Ignorance; as the Use of the Compass, Gunpowder, and Printing; and by the dullest Nations, as the Germans.

Satires and Personal Writings
Thoughts
On Various Subjects (p. 407)

Thoreau, Henry David
Our inventions are wont to be pretty toys, which distract our attention from serious things. They are but improved means to an unimproved end.

Walden
Economy

Twain, Mark
A man invents a thing which could revolutionize the arts, produce mountains of money, and bless the earth, and who will bother with it or show any interest in it?—and so you are just as poor as you were before. But you invent some worthless thing to amuse yourself with, and would throw it away if let alone, and all of a sudden the whole world makes a snatch for it and out crops a fortune.

The American Claimant
Chapter XXIV (p. 228)

Unknown
Necessity is the mother of invention. A race of inventors has sprung up in this country because they were needed. Human labor was scarce and high. A new country was to be conquered and brought under cultivation. Wide fields demanded rapid means of sowing and harvesting. A scanty population and distant markets demanded greater facilities for rapid transit. A high ideal of life demanded a thousand new elements of gratification; and to supply all these demands a thousand new machines and processes had to be invented.

Scientific American
The Best Way to Encourage Invention (p. 192)
Volume XXXVIII, 30 March 1878

Whitehead, Alfred North
The greatest invention of the nineteenth century was the invention of the method of invention.

Science and the Modern World
The Nineteenth Century (p. 96)

Inventive genius requires mental activity as a condition for its vigorous exercise. "Necessity is the mother of invention" is a silly proverb. "Necessity is the mother of futile dodges" is much nearer to the truth. The basis of the growth of modern invention is science, and science is almost wholly the outgrowth of pleasurable intellectual curiosity.

The Aims of Education (p. 69)

According to "Turner's Law", the invocation of the tooth fairy should not occur more than once in any scientific argument.
Marcia Bartusiak – (See p. 173)

INVENTOR

Emerson, Ralph Waldo
Man is a shrewd inventor, and is ever taking the hint of a new machine from his own structure, adapting some secret of his own anatomy in iron, wood, and leather, to some required function in the work of the world.

English Traits
Wealth (p. 83)

Milton, John
Th' invention all admir'd, and each how he
To be th' inventor miss'd; so easy it seem'd,
Once found, which yet unfound most would have thought
Impossible!

Paradise Lost
Book VI, l. 498–501

Rabinow, Jacob
The job of the inventors is to provide the lead for a lagging system.

In Daniel V. DeSimone
Education for Innovation
The Process of Invention (p. 75)

This is the penalty of being an inventor. If you invent something when everybody wants it, it is too late; it's been thought of by everybody else. If you invent too early, nobody wants it because it is too early. If you invent very late, after the need has passed, then it is just a mental exercise. I assure you that it is very hard to invent just at the right time.

In Daniel V. DeSimone
Education for Innovation
The Process of Invention (p. 75)

Redfield, Casper L.
A man's capacity as an inventor depends upon his faculty of making guesses which have some semblance of possibility . . .

In Joseph Rossman
Industrial Creativity: The Psychology of the Inventor (p. 111)

Rossman, Joseph
Inventors are unconscious social changers.
Industrial Creativity: The Psychology of the Inventor (p. 6)

The inventor experiences a need which he wishes to satisfy.
Industrial Creativity: The Psychology of the Inventor (p. 81)

Swann, W.F.G.
The inventor walks in the territory which the man of science has mapped out into regions of assured fertility, dubious fertility, and almost certain sterility. The man of science, and indeed the engineer, are inclined to conserve their efforts by walking in the rather limited realm which, on the basis of the laws with which they operate, represent regions of assured fertility. However, the inventor walks with courage everywhere. He sees a pasture which he thinks has promise. The physicist would explain to him that his reasons for expecting something from that region are invalid, and in 90 per cent of the cases they are, but the inventor walks nevertheless.

In Lenox R. Lohr
Centennial of Engineering
The Engineer and the Scientist (pp. 260–1)

Taylor, Calvin W.
. . . practically all of the mighty rivers of industry spring from the headwaters of lone wolf inventors or creators.

In Daniel V. DeSimone
Education for Innovation
Factors Influencing Creativity (p. 49)

Taylor, Isaac
The great inventor is one who has walked forth upon the industrial world, not from universities, but from hovels; not as clad in silks and decked with honours, but as clad in fustian and grimed with soot and oil.

In Tyron Edwards
The New Dictionary of Thoughts

Twain, Mark
An inventor is a poet—a true poet—and nothing in any degree less than a high order of poet—wherefore his noblest pleasure dies with the stroke that completes the creature of his genius, just as the painter's & the sculptor's & other poets' highest pleasure ceases with the touch that finishes their work—& so only he can understand or appreciate the legitimate "success" of his achievement, little minds being able to get no higher than a comprehension of a vulgar moneyed success.

Letter to Pamela Moffett
6/12/1870

INVESTIGATOR

Cannon, W.B.

The investigator may be made to dwell in a garret, he may be forced to live on crusts and wear dilapidated clothes, he may be deprived of social recognition, but if he has time, he can steadfastly devote himself to research. Take away his free time and he is utterly destroyed as a contributor to knowledge.

The Way of An Investigator (p. 87)

JUDGMENT

Drinker, Henry
. . . in regard to all civil engineering, the first requisite is good judgment, the second requisite is *good judgment*, and the final requisite is GOOD JUDGMENT.

Tunneling, Explosive Compounds, and Rock Drilling (p. 1005)

Atomic energy is the most powerful force known to man—except woman.
Evan Esar – (See p. 55)

KNOWLEDGE

Bacon, Francis
. . . that knowledge hath in it somewhat of the serpent, and therefore where it entereth into a man it makes him swell; *"scientia inflat"* [knowledge puffs up].

Advancement of Learning
First Book, Chapter I, 2

Billings, Josh
It is better to kno less, than to kno so mutch, that aint so.

Old Probability: Perhaps Rain—Perhaps Not

Borel, Émile
Incomplete knowledge must be considered as perfectly normal in probability theory; we might even say that, if we knew all the circumstances of a phenomenon, there would be no place for probability, and we would know the outcome with certainty.

Probability and Certainty
Chapter 1 (p. 13)

Byron, Lord George Gordon
That knowledge is not happiness, and science
But an exchange of ignorance for that
Which is another kind of ignorance.

The Poetical Works of Lord Byron
Manfred, a dramatic poem
Act 2, scene V, l. 431–433

Clarke, J.M.
Knowledge is the only instrument of production that is not subject to diminishing returns.

Journal of Political Economy
Overhead Costs in Modern Industry
October 1927

Collingwood, Robin George
Questioning is the cutting edge of knowledge; assertion is the dead weight behind the edge that gives it driving force.

Speculum Mentis (p. 78)

Cowper, William
Knowledge and Wisdom, far from being one,
Have oft-times no connexion. Knowledge dwells
In heads replete with thoughts of other men;
Wisdom in minds attentive to their own.
Knowledge, a rude unprofitable mass,
The mere materials with which wisdom builds . . .
Knowledge is proud that he has learn'd so much;
Wisdom is humble that he knows no more.

The Complete Poetical Works of William Cowper
The Task
Book VI, l. 88–93, 96, 97

da Vinci, Leonardo
The acquisition of any knowledge whatever is always useful to the intellect, because it will be able to banish the useless things and retain those which are good. For nothing can be either loved or hated unless it is first known.

The Notebooks of Leonardo da Vinci
Volume I
Aphorisms (p. 88)

Dickens, Charles
But wot's that, you're a-doin' of? Pursuit of knowledge under difficulties, Sammy?

The Pickwick Papers
Chapter 33 (p. 456)

Einstein, Albert
Yet it is equally clear that knowledge of what *is*, does not open the door directly to what *should be*.

Out of My Later Years (p. 22)

Emerson, Ralph Waldo
All our progress is an unfolding, like the vegetable bud. You have first an instinct, then an opinion, then a knowledge.

Essays
Intellect

Our knowledge is the amassed thought and experience of innumerable minds.

Letters and Social Aims
Quotations and Originality (p. 177)

Fabing, Harold
Marr, Ray
Knowledge is a process of piling up facts; wisdom lies in their simplification.

Fischerisms (p. 2)

Fuller, Thomas
Knowledge is a Treasure, but Practice is the Key to it.

Gnomologia
Number 3139

Gibran, Kahlil
Knowledge and understanding are life's faithful companions who will never prove untrue to you. For knowledge is your crown, and understanding your staff; and when they are with you, you possess no greater treasures.

The Treasured Writing of Kahlil Gibran
The Voice of the Master
Wisdom
viii (p. 488)

Holmes, Oliver Wendell
It is the province of knowledge to speak and it is the privilege of wisdom to listen.

The Poet at the Breakfast-Table
Chapter 10

Scientific knowledge, even in the most modest persons, has mingled with it a something which partakes of insolence.

The Autocrat of the Breakfast-Table
Chapter 3

Knowledge and timber shouldn't be much used till they are seasoned.

The Autocrat of the Breakfast-Table
Chapter 6

Huxley, Aldous
Knowledge is power and, by a seeming paradox, it is through their knowledge of what happens in this unexperienced world of abstractions and inferences that scientists and technologists have acquired their enormous and growing power to control, direct and modify the world of manifold appearances in which human beings are privileged and condemned to live.

Literature and Science
Chapter 3 (p. 9)

Knowledge is the highest good, truth the supreme value, all the rest is secondary and subordinate.

Brave New World (pp. 178–9)

Huxley, Thomas H.
Indeed, if a little knowledge is dangerous, where is the man who has so much as to be out of danger?

Collected Essays
Volume III
On Elementary Instruction in Physiology

Jefferson, Thomas
A patient pursuit of facts, and cautious combination and comparison of them, is the drudgery to which man is subjected by his Master, if he wishes to attain sure knowledge.

Notes on the State of Virginia
Chapter 6 (p. 72, n)

Jevons, W.S.
I am convinced that it is impossible to expound the methods of induction in a sound manner, without resting them on the theory of probability. Perfect knowledge alone can give certainty, and in nature perfect knowledge would be infinite knowledge, which is clearly beyond our capacities. We have, therefore, to content ourselves with partial knowledge,—knowledge mingled with ignorance, producing doubt.

The Principles of Science
Chapter X (p. 197)

Johnson, Samuel
Integrity without knowledge is weak and useless and knowledge without integrity is dangerous and dreadful.

Rasselas
Chapter 41 (p. 317)

Lewis, Clarence Irving
Interest in truth for its own sake—the pure and undistracted purpose to know—is not the characteristic final purpose of knowing. Knowledge for its own sake, and the contemplative life, represent an esthetic or near-esthetic ideal rather than one normally attributable to cognition. It is merely a professional fallacy of the scholar to impute his own peculiar interest in finding out the truth to human cognizing in general, as if that were the aim which rules or should rule it. He who is disinterestedly interested in finding out and knowing; who subordinates the desires and interests of action to discovery of truth, and to contemplation of it; likewise divests knowledge of its natural and pragmatic significance. By the same token, the ideal of the contemplative life is mildly abnormal,

however valid and indubitable the values to which it is addressed. The Ivory tower is characteristically the refuge of the practically defeated and of those who become disillusioned of the utilities of action.

An Analysis of Knowledge and Valuation
Chapter XIV (p. 442)

Lichtenberg, Georg Christoph
I made the journey to knowledge like dogs who go for walks with their masters, a hundred times forward and backward over the same territory; and when I arrived I was tired.

Lichtenberg: Aphorisms & Letters
Aphorisms (p. 58)

Marlowe, Christopher
Our souls, whose faculties can comprehend
The wondrous architecture of the world,
And measure every wandering planet's course,
Still climbing after knowledge infinite.

Tamburlaine the Great
Part the First
Act II, scene 7, l. 20–23

Myrdal, Gunnar
All ignorance, like all knowledge, tends thus to be opportunist.

Objectivity in Social Research
Chapter III (p. 19)

Stewart, Ian
I may not understand it, but it sure looks important to me.

Does God Play Dice? (p. 121)

Szent-Györgyi, Albert
Knowledge is a sacred cow, and my problem will be how we can milk her while keeping clear of her horns.

Science
Teaching and the Expanding Knowledge (p. 1278)
Volume 146, Number 3649, 4 December 1964

Tennyson, Alfred Lord
Knowledge comes, but wisdom lingers.

The Complete Poetical Works of Tennyson
Locksley Hall
l. 141

LAWS

Bartusiak, Marcia
According to "Turner's Law", the invocation of the tooth fairy should not occur more than once in any scientific argument.

Thursday's Universe (p. 207)

Dewey, John
Scientific principles and laws do not lie on the surface of nature. They are hidden, and must be wrested from nature by an active and elaborate technique of inquiry.

Reconstruction in Philosophy (p. 32)

Euclid
The laws of nature are but the mathematical thoughts of God.

In Stanley Gudder
A Mathematical Journey (p. 112)

Feynman, Richard P.
There is also a rhythm and a pattern between the phenomena of nature which is not apparent to the eye, but only to the eye of analysis; and it is these rhythms and patterns which we call Physical Laws.

The Character of Physical Law
Chapter 1 (p. 13)

Huxley, Thomas H.
The chess-board is the world; the pieces are the phenomena of the universe; the rules of the game are what we call the laws of Nature. The player on the other side is hidden from us. We know that his play is always fair, just, and patient. But also we know, to our cost, that he never overlooks a mistake, or makes the smallest allowance for ignorance.

Lay Sermons, Addresses, and Reviews
A Liberal Education (pp. 31–2)

Laplace, Pierre Simon
All the effects of nature are only the mathematical consequences of a small number of immutable laws.

In E.T. Bell
Men of Mathematics (p. 172)

Pearson, Karl
Scientific Law is description, not a prescription.

In Henry Crew
General Physics (p. 45)

Planck, Max
Thus from the outset we can be quite clear about one very important fact, namely, that the validity of the law of causation for the world of reality is a question that cannot be decided on grounds of abstract reasoning.

Where is Science Going? (p. 113)

Wheeler, John A.
There is no law except that there is no law.

In John D. Barrow
The World within the World (p. 293)

LEVER

Archimedes
Give me a place on which to stand and I will move the world.

In Pappus of Alexandria
Collection

Friend of Archimedes
Give me a place to sit, and I'll watch.

Graham, L.A.
See Saw, Marjorie Daw,
She rocked—and learned the lever law.
She saw that he weighed more than she
For she sat higher up than he,
Which made her cry, excitedly,
"To see saw, it is plain to see,
There must be an equality
'Twixt You times X and Y times Me."

Ingenious Mathematical Problems and Methods
Mathematical Nursery Rhyme Number 19

LOGIC

Boutroux, Pierre
Logic is invincible because in order to combat logic it is necessary to use logic.

In Morris Kline
Mathematical Thought from Ancient to Modern Times (p. 1182)

Carroll, Lewis
"I know what you're thinking about," said Tweedledum; "but it isn't so, nohow."

"Contrariwise," continued Tweedledee, "if it was so, it might be; and if it were so, it would be; but as it isn't, it ain't. That's logic."

The Complete Works of Lewis Carroll
Through the Looking Glass
Tweedledum and Tweedledee

Chesterton, Gilbert Keith
A great deal is said in these days about the value or valuelessness of logic. In the main, indeed, logic is not a productive tool so much as a weapon for defence. A man building up an intellectual system has to build like Nehemiah, with the sword in one hand and the trowel in the other. The imagination, the constructive quality, is the trowel, and the argument is the sword. A wide experience of actual intellectual affairs will lead most people to the conclusion that logic is mainly valuable as a weapon wherewith to exterminate logicians.

The G.K. Chesterton Calendar
January Ten

Clough, Arthur Hugh
Good, too, Logic, of course; in itself, but not in fine weather.

In James R. Newman
The World of Mathematics
Volume Three
The Bothie of Tober-na-Vuolich (p. 1878)

Colton, Charles Caleb
Logic is a large drawer, containing some useful instruments, and many
more that are superfluous. A wise man will look into it for two purposes,
to avail himself of those instruments that are really useful, and to admire
the ingenuity with which those that are not so, are assorted and arranged.

Lacon (p. 163)

Dunsany, Lord Edward John Moreton Drax Plunkett
But, logic, like whiskey, loses its
beneficial effect when taken in too
large quantities.

My Ireland
Weeds & Moss (p. 159)

I THINK THAT'S QUITE ENOUGH
LOGIC FOR ONE EVENING..!

Heaviside, Oliver
Logic can be patient for it is eternal.

In Morris Kline
*Mathematical Thought from Ancient to
Modern Times* (p. 3)

Holmes, Oliver Wendell
Logic is logic. That's all I say.

The Autocrat of the Breakfast-Table
Chapter 11

Hubbard, Elbert
Logic is one thing and common sense
another.

The Note Book (1927)

Jerome, Jerome K.
When a twelfth century youth fell in love, he did not take three paces
backward, gaze into her eyes, and tell her she was too beautiful to live.
He said he would step outside and see about it. And if, when he got
out, he met a man and broke his head—the other man's head, I mean—
then that proved that his—the first fellow's girl—was a pretty girl. But
if the other fellow broke his head—not his own you know, but the other
fellow's—the other fellow to the second fellow, that is, because of course
the other fellow would only be the other fellow to him, not the first fellow
who—well, if he broke his head, then his girl—not the other fellow's, but
the fellow who was the—Look here, if A broke B's head, then A's girl
was a pretty girl: but if B broke A's head, then A's girl wasn't a pretty
girl, but B's girl was.

The Idle Thoughts of an Idle Fellow
On Being Idle

Jones, Raymond F.
Logic hasn't wholly dispelled the society of witches and prophets and sorcerers and soothsayers.

The Non-Statistical Man (p. 85)

Jowett, Benjamin
Logic is neither a science nor an art, but a dodge.

In James R. Newman
The World of Mathematics
Volume Four (p. 2402)

Russell, Bertrand
. . . logic is the youth of mathematics . . .

Introduction to Mathematical Philosophy
Mathematics and Logic (p. 194)

. . . the rules of logic are to mathematics what those of structure are to architecture.

Philosophical Essays
The Study of Mathematics (p. 74)

Schiller, F.C.S.
Among the obstacles to scientific progress a high place must certainly be assigned to the analysis of scientific procedure which Logic has provided . . . It has not tried to describe the methods by which the sciences have actually advanced, and to extract . . . rules which might be used to regulate scientific progress, but has freely rearranged the actual procedure in accordance with its prejudices. For the order of discovery there has been substituted an order of 'proof' . . .

In Charles Singer (Editor)
Studies in the History and Method of Science
Scientific Discovery and Logical Proof (p. 235)

. . . it is not too much to say that the more deference men of science have paid to Logic, the worse it has been for the scientific value of their reasoning . . . Fortunately for the world, however, the great men of science have usually been kept in salutary ignorance of the logical tradition . . .

In Charles Singer (Editor)
Studies in the History and Method of Science
Scientific Discovery and Logical Proof (p. 236)

Tagore, Rabindranath
A mind all logic is like a knife all blade,
It makes the hand bleed that uses it!

Collected Poems and Plays of Rabindranath Tagore (p. 312)

Unknown
Reiteration of an argument is often more effective than its inherent logic.

Source unknown

Whitehead, Alfred North
Logic, properly used, does not shackle thought. It gives freedom, and above all, boldness. Illogical thought hesitates to draw conclusions, because it never knows either what it means, or what it assumes, or how far it trusts its own assumptions, or what will be the effect of any modification of assumptions.

The Organization of Thought (p. 132)

Neither logic without observation, nor observation without logic, can move one step in the formation of science.

The Organization of Thought (p. 132)

Wittgenstein, Ludwig
In der Logik ist nichts zufällig.
[Nothing, in logic, is accidental.]

Tractatus Logico Philosophicus
2.012 (p. 31)

The disastrous invasion of mathematics by logic.

Remarks on the Foundations of Mathematics (p. 145)

MACHINE

Boorstin, Daniel
Just as the American's love affair with his land produced pioneering adventures and unceasing excitement in the conquest of the continent, so too his latter-day romance with the Machine produced pioneering adventures—of a new kind . . . there were no boundaries to a machine-made world.

Hidden History
From the Land to the Machine (pp. 252–3)

Bottomley, Gordon
Your worship is your furnaces,
Which, like old idols, lost obscenes,
Have molton bowels; your vision is
Machines for making more machines.

Poems of Thirty Years
To Iron-Founders and Others

Chesterton, Gilbert Keith
A machine *is* a machine because it cannot think.

The G.K. Chesterton Calendar
February Sixteen

de Saint-Exupéry, Antoine
. . . the machine does not isolate man from the great problems of nature but plunges him more deeply into them.

Wind, Sand and Stars (p. 67)

Disraeli, Benjamin
A machine is a slave that neither brings nor bears degradation.

Coningsby
Book IV, Chapter 2 (p. 179)

. . . the mystery of mysteries is to view machines making machines . . .

Coningsby
Book IV, Chapter 2 (p. 180)

Florman, Samuel C.
In his emotional involvement with the machine, the engineer cannot help but feel at times that he has come face to face with a strange but potent form of life.

The Existential Pleasures of Engineering (p. 139)

Goodman, Ellen
Once upon a time we were just plain people. But that was before we began having relationships with mechanical systems. Get involved with a machine and sooner or later you are reduced to a factor.

Washington Post
The Human Factor
January 1987

Lee, Gerald Stanley
It is never the machines that are dead.
It is only the mechanically-minded men that are dead.

Crowds
Book III, Part II, Chapter V (p. 249)

Lippmann, Walter
You cannot endow even the best machine with initiative: the jolliest steamroller will not plant flowers.

A Preface to Politics
Routineer and Inventor (p. 30)

Platonov, Andrei
Frossia's husband had the ability to feel the voltage of an electric current like a personal emotion. He animated everything that his hands or mind touched, so he really understood the flow of forces in any piece of mechanism and could actually feel the painful, patient resistance of the metal body of a machine.

Fro and Other Stories
Fro (p. 88)

Russell, Bertrand
Machines are worshipped because they are beautiful, and valued because they confer power; they are hated because they are hideous, and loathed because they impose slavery.

Sceptical Essays
Chapter VI (p. 83)

A machine is like a Djinn in the Arabian Nights: beautiful and beneficent to its master, but hideous and terrible to his enemies.

Sceptical Essays
Chapter VI (p. 83)

Ryle, Gilbert
. . . the dogma of the Ghost in the Machine.

The Concept of Mind (pp. 15–16)

Samuel Arthur L.
A machine is not a genie, it does not work by magic, it does not possess
a will, and Wiener to the contrary, nothing comes out which has not
been put in, barring of course, an infrequent case of malfunctioning
. . . The "intentions" which the machine seems to manifest are the
intentions of the human programmer, as specified in advance, or they
are subsidiary intentions derived from these, following rules specified
by the programmer . . . the machine *will not* and *cannot* do any of these
things until it has been instructed as to how to proceed . . . To believe
otherwise is either to believe in magic or to believe that the existence of
man's will is an illusion and that man's actions are as mechanical as the
machine's.

Science
Some Moral and Technical Consequences of Automation—A Refutation (p. 741)
Volume 132, Number 3429, 16 September 1960

Schumacher, E.F.
Ever bigger machines, entailing ever bigger concentrations of economic
power and exerting ever greater violence against the environment, do
not represent progress: they are a denial of wisdom. Wisdom demands
a new orientation of science and technology towards the organic, the
gentle, the non-violent, the elegant and beautiful.

Small is Beautiful
Part I, Chapter II (p. 31)

MATHEMATICS

Bacon, Roger
For the things of this world cannot be made known without a knowledge of mathematics.

Opus Majus
Part 4, Distinctia Prima, cap 1, 1267

Bellman, Richard
Mathematics makes natural questions precise.

Eye of the Hurricane (p. 114)

Buchanan, Scott
The prestige of the engineer is another accretion to the tradition of mathematics. This more than any other one thing accounts for our present mathematical complex. The engineer is fast taking the position of authority, superseding the priest, the scholar, and the statesman in our organized thought and action.

Poetry and Mathematics (pp. 37–8)

Bullock, James
Mathematics is not a way of hanging numbers on things so that quantitative answers to ordinary questions can be obtained. It is a language that allows one to think about extraordinary questions.

American Mathematical Monthly
Literacy in the Language of Mathematics (p. 737)
Volume 101, Number 8, October 1994

Crick, Francis
Mathematics cares neither for science nor for engineering (except as a source of problems) but only about the relationship between abstract entities.

What Mad Pursuit (p. 160)

Cross, Hardy
There is an unfortunate tendency to burden engineers, through books, with endless techniques and procedures of mathematical analysis. Few students know that at best books can furnish only a perishable net of large mesh through which they may begin to strain their information and that every fiber of that net must be rewoven from man's own thinking and that many new strands must be added if it is to be permanent and reliable in holding the selected data of years of engineering practice. Books present the sets of tools; it is the task of the analytical engineer to select those tools which can be used most advantageously.

Engineers and Ivory Towers
For Man's Use of God's Gifts (p. 106)

Harrington, Eldred
Mathematics is a wonderful tool but it is only a tool; it is not a god to be worshipped.

An Engineer Writes about People and Places and Projects (p. 196)

Kasner, Edward
Newman, James R.
. . . it is to the definite integral that structural engineers must render thanks for the Golden Gate Bridge, for it rests on this even more than on concrete and steel . . .

Mathematics and the Imagination (p. 340)

Lieber, Lillian R.
When we learn to drive a car we are able to "go places" easily and pleasantly instead of walking to them with a great deal of effort. And so you will see that the more Mathematics we know the EASIER life becomes, for it is a TOOL with which we can accomplish things that we could not do at all with our bare hands. Thus Mathematics helps our brains and hands and feet, and can make a race of supermen out of us.

The Education of T.C. MITS (p. 45)

Lobachevskii, Nikolai Ivanovich
There is no branch of mathematics however abstract which may not some day be applied to phenomena of the real world.

In Stanley Gudder
A Mathematical Journey (p. 36)

Nordenholt, George F.
Smart creative workers are those who are quick to see the limitations of mathematical calculations.

Product Engineering
A Graduate Can Measure A Bottle
Editorial
April 1953

Page, Ray
. . . Today our world of automation revolves around science and science in turn rests on mathematics.

Quote (p. 380)
Volume 53, Number 20, 14 May 1967

Poincaré, Henri
The engineer should receive a complete mathematical education, but for what should it serve him?

To see the different aspects of things and to see them quickly; he has no time to hunt mice.

The Foundations of Science
Science and Method (p. 438)

Schlicter, Dean
Go down deep enough into anything and you will find mathematics.

The Mathematics Teacher
In Margaret Joseph
The Future of Geometry (p. 29)
January 1936, Volume XXIX, Number 1

Spengler, Oswald
Gothic cathedrals and Doric temples are *mathematics in stone*.

The Decline of the West
Volume I, Chapter II, section ii (p. 58)

Spiegel, M.R.
An engineer called E has students who complain
That mathematics taught to them is nothing but a pain.

"Why can't we take a course," they ask, "a course designed for us?
A course which helps us in our fields without this kind of fuss.

Why ask us for a proof and make us say, 'I can't, sir,'
When what we'd really like to know is how to get the answer."

With empathy E rushed to M who gave the ill-famed course
And told him of the problem and also of the source.

"The students say the course you give drives them up the wall.
Epsilons and deltas seem to have no point at all.

They need the kind of math that helps with circuitry and cables.
They need to know how they can use the mathematics tables."

Professor M was not surprised—he'd heard complaints before.
"You realize," he said to E, "there are things we can't ignore.

Math requires subtleties—you can't just make a list.
You cannot take derivatives of things that don't exist."

On and on he lectured E with little hesitation
On topics such as limits and improper integration,

On necessary and sufficient conditions for existence.
E tried, but could not shake, Professor M's insistence.

There is no end to this debate; E and M cannot agree.
There are two sides and each will teach as only he can see.

Mathematics Magazine
E and M (p. 139)
Volume 54, Number 3, May 1981

Unknown
The first law of Engineering Mathematics: All infinite series converge,
and moreover converge to the first term.

Source unknown

Wright, Frank Lloyd
. . . mathematics in co-ordinated Form is architecture.

Frank Lloyd Wright: An Autobiography
Hollyhock House in Hollywood (p. 227)

MEASUREMENT

Asimov, Isaac
. . . we must remember that measures were made for man and not man for measures.

Of Time and Space and Other Things

Beyer, Robert W.

10^{-6} phones	= 1 microphone
10^{-2} pedes	= 1 centipede
10^6 phones	= 1 megaphone
10^{-12} boos	= 1 picoboo
10^{-18} boys	= 1 attoboy
10^{-2} mental	= 1 centimental
10^{-1} cards	= 1 decacards
10^{12} bulls	= 1 terrabul
2 gorics	= 1 paragoric
10^9 los	= 1 gigolo
10^{-1} mates	= 1 decimate
10^{-3} cans	= 1 millican
2×10^2 with its	= 2 hecto with it
10^1 dents	= 1 decadent
10^{-3} Nanetts	= 1 nonoo Nanette
2×10^3 mockingbirds	= 2 kilomockingbird
10^{-3} taries	= 1 military
10^3 monjaros	= 1 kilomanjara
10^{12} fermis	= 1 terra fermi
10^{-6} fish	= 1 microfishe

The Physics Teacher
Humor Rumor: it helps! (p. 575)
December 1977

Bondi, Hermann
A quantity like time, or any other physical measurement, does not exist in a completely abstract way. We find no sense in talking about something unless we specify how we measure it. It is the definition by the method of measuring a quantity that is the one sure way of avoiding talking nonsense . . .

Relativity and Common Sense
Chapter VII (p. 65)

Chaucer, Geoffrey
In everything there lieth measure.

Troylus and Cryseyde
c

Deming, William Edwards
It is important to realize that it is not the one measurement, alone, but its relation to the rest of the sequence that is of interest.

Statistical Adjustment of Data (p. 3)

Dewey, John
Insistence upon numerical measurement when it is not inherently required by the consequence to be effected, is a mark of respect for the ritual of scientific practice at the expense of its substance.

Logic: The Theory of Inquiry
Chapter XI (p. 205)

Fox, Russell
Gorbuny, Max
Hooke, Robert
Measurement has meaning only if we can transmit the information without ambiguity to others.

The Science of Science
Part II, 4 (p. 31)

Isaiah 40:12
Who was it measured the waters of the sea in the hollow of his hand and calculated the dimensions of the heavens, gauged the whole earth to the bushel, weighed the mountains in scales, the hills in a balance?

The Bible

Kaplan, Abraham
Measurement, we have seen, always has an element of error in it. The most exact description or prediction that a scientist can make is still only approximate. If, as sometimes happens, a perfect correspondence with observation does appear, it must be regarded as accidental, and, as

Jevons [see *The Principles of Science*, p. 457] . . . remarks, it "should give rise to suspicion rather than to satisfaction."

The Conduct of Inquiry
Chapter VI, section 25 (p. 215)

Proleptically, I would say that whether we can measure something depends, not on that thing, but on how we have conceptualized it, on our knowledge of it, above all on the skill and ingenuity which we can bring to bear on the process of measurement which our inquiry can put to use.

The Conduct of Inquiry
Chapter V, section 20 (p. 176)

Krutch, Joseph Wood

We are committed to the scientific method and measurement is the foundation of that method; hence we are prone to assume that whatever is measurable must be significant and that whatever cannot be measured may as well be disregarded.

Human Nature and the Human Condition
Chapter 5 (p. 78)

Lewis, G.N.

I have no patience with attempts to identify science with measurement, which is but one of its tools, or with any definition of the scientist which would exclude a Darwin, a Pasteur, or a Kekulé.

The Anatomy of Science (p. 6)

Peter, Lawrence J.

Coomb's Law. If you can't measure it, I'm not interested.

Human Behavior
Peter's People (p. 9)
August 1976

Reynolds, H.T.

Crude measurement usually yields misleading, even erroneous conclusions no matter how sophisticated a technique is used.

Analysis of Nominal Data (p. 56)

Whitehead, Alfred North

Consider . . . the scientific notion of measurement. Can we elucidate the turmoil of Europe by weighing its dictators, its prime ministers, and its editors of newspapers? The idea is absurd, although some relevant information might be obtained. I am not upholding the irrelevance of science. Such a doctrine would be foolish. For example, a daily record of the bodily temperature of the men, above mentioned, might be useful. My point is the incompleteness of the information.

Modes of Thought (p. 25–6)
Creative impulse

Unknown

What the measurements will not do, is to get you out of the crisis you are already in.

Source unknown

Chocolate Chip Cookies:

Ingredients:

1. 532.35 cm^3 gluten
2. 4.9 cm^3 $NaHCO_3$
3. 4.9 cm^3 refined halite
4. 236.6 cm^3 partially hydrogenated tallow triglyceride
5. 177.45 cm^3 crystalline $C_{12}H_{22}O_{11}$
6. 177.45 cm^3 unrefined $C_{12}H_{22}O_{11}$
7. 4.9 cm^3 methyl ether of protocatechuic aldehyde
8. Two calcium carbonate-encapsulated avian albumen-coated protein
9. 473.2 cm^3 theobroma cacao
10. 236.6 cm^3 de-encapsulated legume meats (sieve size #10)

To a 2-L jacketed round reactor vessel (reactor #1) with an overall heat transfer coefficient of about 100 Btu/F-ft^2-hr, add ingredients one, two and three with constant agitation. In a second 2-L reactor vessel with a radial flow impeller operating at 100 rpm, add ingredients four, five, six and seven until the mixture is homogenous. To reactor #2, add ingredient eight, followed by three equal volumes of the homogenous mixture in reactor #1. Additionally, add ingredients nine and ten slowly, with constant agitation. Care must be taken at this point in the reaction to control any temperature rise that may be the result of an exothermic reaction.

Using a screw extrude attached to a #4 nodulizer, place the mixture piece-meal on a 316SS sheet (300 × 600 mm). Heat in a 460 K oven for a period of time that is in agreement with Frank & Johnston's first order rate expression (see JACOS, 21, 55), or until golden brown. Once the reaction is complete, place the sheet on a 25C heat-transfer table, allowing the product to come to equilibrium.

Source unknown

MECHANICS

Cross, Hardy
Mechanics, for instance, is a diamond of many facets and scintillates with different colors for the mathematician, the student of pure physics, the student of cosmic physics or the engineer.

Engineers and Ivory Towers
The Education of an Engineer (p. 55)

da Vinci, Leonardo
Mechanics is the paradise of the mathematical sciences because by means of it one comes to the fruit of mathematics.

The Notebooks of Leonardo da Vinci
Volume I
Mathematics (p. 628)

Gregory, Olinthus
The science of mechanics, whether considered in its theory as a subject of curious and refined speculations, calculated for the learned, ingenious, and contemplative, or in practice as contribution to the conveniences and elegancies of life, and the wealth of nations, may be ranked the first and most important of all human acquirements.

American Journal of Science
A Treatise of Mechanics (p. 72)
Volume 7, 1824

Osgood, W.F.
Mechanics is a natural science and like any natural science requires for its comprehension the observation and knowledge of a vast fund of individual cases . . .

But Mechanics is not an empirical subject in the sense in which physics and chemistry, when dealing with the border region of the human knowledge of the day, are empirical . . . The laws of Mechanics, like the laws of Geometry, so far as first approximations go—the laws that explain the motion of the golf ball or the gyroscope or the skidding automobile, and which make possible the calculation of lunar tables and the prediction of eclipses—these laws are known, and will be as new and important two thousand years hence, as in the recent past of science when first they emerged into the light of day . . .

The world in which the boy and girl have lived is the true laboratory of elementary mechanics. The tennis ball, the golf ball, the shell on the river; the automobile—good old Model T, in its day, and the home-made autos and motor boats which youngsters construct and will continue to construct—the amateur printing press; the games in which the mechanics of the body is a part; all these things go to provide the student with rich laboratory experiences . . .

Mechanics (pp. v–vi)

METHOD

Camus, Albert
When one has no character one *has* to apply a method.

The Fall (p. 11)

Cohen, Morris R.
Nagel, Ernest
. . . the safety of science depends on there being men who care more for the justice of their methods than for any results obtained by their use.

An Introduction to Logic and Scientific Method
Chapter XX, section 2 (p. 402)

Hilbert, David
He who seeks for methods without having a definite problem in mind seeks for the most part in vain.

Bulletin American Mathematical Society
Mathematical Problems (p. 444)
Volume 8

Holmes, Sherlock
Pon my word Watson, you are coming along wonderfully. We have really done very well indeed. It is true that you have missed everything of importance, but you have hit upon the method.

In Arthur Conan Doyle
The Complete Sherlock Holmes
A Case of Identity

Pólya, George
My method to overcome a difficulty is to go round it.

How to Solve It (p. 181)

MISTAKE

Gombrich, E.H.

In order to learn, we must make mistakes, and the most fruitful mistakes which nature could have implanted in us would be the assumption of even greater simplicities than we are likely to meet in this bewildering world of ours . . . To probe a hole we first use a straight stick to see how far it takes us. To probe the visible world we use the assumption that things are simple until they prove to be otherwise.

In John Pottage
Geometrical Investigations (p. 15)

Siegel, Eli

If a mistake is not a stepping stone, it is a mistake.

Damned Welcome (p. 39)

MODEL

Eigen, Manfred
A theory has only the alternative of being right or wrong. A model has a third possibility: it may be right, but irrelevant.

In Jagdish Mehra (Editor)
The Physicist's Conception of Nature (p. 618)

Kaplan, Abraham
The words "model" and "mode" have, indeed, the same root; today, model building is science *à la mode*.

The Conduct of Inquiry
Chapter VII, section 30 (p. 258)

Karlin, Samuel
The purpose of models is not to fit the data but to sharpen the questions.

11th R.A. Fisher Memorial Lecture
Royal Society
20 April 1983

Kelvin, William Thomson, Baron
I never satisfy myself until I can make a mechanical model of a thing. If I can make a mechanical model, I understand it.

Baltimore Lectures on Molecular Dynamics, and the Wave Theory of Light (p. 270)

Unknown
The sciences do not try to explain, they hardly even try to interpret, they mainly make models.

Source unknown

MOTION

Cohen, I. Bernard
Odd as it may seem, most people's views about motion are part of a system of physics that was proposed more than 2,000 years ago and was experimentally shown to be inadequate at least 1,400 years ago.

The Birth of A New Physics (p. 3)

Galilei, Galileo
My purpose is to set forth a very new science dealing with a very ancient subject. There is, in nature, perhaps nothing older than motion, concerning which the books written by philosophers are neither very few nor small; nevertheless, I have discovered by experiment some properties of it which are worth knowing and which have not hitherto been either observed or demonstrated.

Dialogues Concerning Two New Sciences
Third Day
Change of Position

Jeans, Sir James Hopwood
. . . the laws which nature obeys are less suggestive of those which a machine obeys in its motion than of those which a musician obeys in writing a fugue, or a poet in composing a sonnet. The motions of electrons and atoms do not resemble those of the parts of a locomotive so much as those of the dancers in a cotillion. And if the "true essence of substances" is for ever unknowable, it does not matter whether the cotillion is danced at a ball in real life, or on a cinematography screen, or in a story of Boccaccio.

The Mysterious Universe
Into Deep Waters (p. 136)

Wittgenstein, Ludwig

The fact that we can describe the motions of the world using Newtonian mechanics tells us nothing about the world. The fact that we do, does tell us something about the world.

In John D. Barrow
The World within the World (p. 77)

You cannot endow even the best machine with initiative: the jolliest steamroller will not plant flowers.
Walter Lippmann – (See p. 181)

OBSERVATION

Anscombe, F.J.
No observations are absolutely trustworthy.

Technometrics
Rejection of Outliers (p. 124)
Volume 2, Number 2, May 1960

Aristotle
. . . while those whom devotion to abstract discussions has rendered unobservant of the facts are too ready to dogmatize on the basis of a few observations.

On Generation and Corruption
Book I, Chapter II

Aurelius, Marcus [Antoninus]
Consider that everything which happens, happens justly, and if thou observest carefully, thou wilt find it to be so.

The Meditations of the Emperor Antoninus Marcus Aurelius
Book IV, section 10

Ayres, C.E.
When Moses emerged from the cloudy obscurity of Mount Sinai and stood before the people with the stone tablets in his hand, he announced that his laws were based on direct observation. It is not recorded that any one doubted him.

Science: The False Messiah (p. 42)

Bachelard, Gaston
A scientific observation is always a committed observation. It conforms or denies one's preconceptions, one's first ideas, one's plan of observation. It shows by demonstration. It structures the phenomena. It transcends what is close at hand. It reconstructs the real after having reconstructed its representation.

Le Nouvel Esprit Scientifique
Volume I

Bernard, Claude
Speaking concretely, when we say "making experiments or making observations," we mean that we devote ourselves to investigation and to research, that we make attempts and trials in order to gain facts from which the mind, through reasoning, may draw knowledge or instruction.

Speaking in the abstract, when we say, "relying on observation and gaining experience," we mean that observation is the mind's support in reasoning, and experience the mind's support in deciding, or still better, the fruit of exact reasoning applied to the interpretation of facts.

Observation, then, is what shows facts; experiment is what teaches about facts and gives experience in relation to anything.
An Introduction to the Study of Experimental Medicine (p. 11)

Blake, William
A fool sees not the same tree that a wise man sees.
The Complete Writings of William Blake
The Marriage of Heaven and Hell
Proverbs of Hell
l. 8

Bohr, Niels
The great extension of our experience in recent years has brought to light the insufficiency of our simple mechanical conceptions and, as a consequence, has shaken the foundation on which the customary interpretation of observation was based . . .
Atomic Theory and Description of Nature
Introductory Survey (p. 2)

Box, G.E.P.
To find out what happens to a system when you interfere with it you have to interfere with it (not just passively observe it).
Technometrics
Use and Abuse of Regression (p. 629)
Volume 8, Number 4, November 1966

Carlyle, Thomas
Shakespeare says, we are creatures that look before and after: the more surprising that we do not look round a little, and see what is passing under our very eyes.
Sartor Resartus
Book I, Chapter 1

Darwin, Charles
Oh, he is a good observer, but he has no power of reasoning!
The Life and Letters of Charles Darwin
Mental Qualities (p. 83)

Eddington, Sir Arthur
For the truth of the conclusions of physical science, observation is the supreme Court of Appeal. It does not follow that every item which we confidently accept as physical knowledge has actually been certified by the Court; our confidence is that it would be certified by the Court if it were submitted. But it does follow that every item of physical knowledge is of a form which might be submitted to the Court. It must be such that we can specify (although it may be impracticable to carry out) an observational procedure which would decide whether it is true or not. Clearly a statement cannot be tested by observation unless it is an assertion about the results of observation. Every item of physical knowledge must therefore be an assertion of what has been or would be the result of carrying out a specified observational procedure.

The Philosophy of Physical Science (pp. 9–10)

Einstein, Albert
A man should look for what is, and not for what he thinks should be . . .

In Peter Michelmore
Einstein: Profile of the Man (p. 20)

It is the theory which decides what we can observe.

In Werner Heisenberg
Physics and Beyond: Encounters and Conversations (p. 77)

Emerson, Ralph Waldo
The difference between landscape and landscape is small but there is a great difference in the beholders.

Essays
Nature

Fabing, Harold
Marr, Ray
You must acquire the ability to describe your observations and your experiences in such language that whoever observes or experiences similarly will be forced to the same conclusion.

Fischerisms (p. 8)

Greer, Scott
. . . the link between observation and formulation is one of the most difficult and crucial in the scientific enterprises. It is the process of interpreting our theory or, as some say, of "operationalizing our concepts". Our creations in the world of possibility must be fitted in the world of probability; in Kant's epigram, "Concepts without precepts are empty". It is also the process of relating our observations to theory; to finish the epigram, "Precepts without concepts are blind".

The Logic of Social Inquiry (p. 160)

Heisenberg, Werner

What we observe is not nature itself, but nature exposed to our method of questioning.

Physics and Philosophy (p. 58)

Holmes, Sherlock

You see, but you do not observe. The distinction is clear.

In Arthur Conan Doyle
The Complete Sherlock Holmes
A Scandal in Bohemia

Hooke, Robert

The truth is, the Science of Nature has been already too long made only a work of the *Brain* and the *Fancy*: It is now high time that it should return to the plainness and soundness of Observations on *material* and *obvious* things.

Micrographia
Preface

Hutten, Ernest H.

. . . certain conditions under which the observable thing is perceived are tacitly assumed . . . for the possibility that we deal with hallucinations or a dream can never be excluded.

The Language of Modern Physics (p. 51)

Isaiah 42:20

You have seen many things but not observed them; your ears are open but you do not hear.

The Bible

Jonson, Ben

I do love to note and to observe.

Volpone
Act II, scene 1

Lyttleton, R.A.

Observations are meaningless without a theory to interpret them.

In Charles-Albert Reichen
A History of Astronomy (p. 88)

O'Neil, W.M.

It urges the scientist, in effect, not to take risks incurred in moving far from the facts. However, it may properly be asked whether science can be undertaken without taking the risks of skating on the possibly thin ice of supposition. The important thing to know is when one is on the more solid ground of observation and when one is on the ice.

Fact and Theory (p. 154)

Poincaré, Henri
It is not sufficient merely to observe; we must use our observations, and
for that purpose we must generalise.

Science and Hypothesis (p. 141)

Pope, Alexander
To observations which ourselves we make,
We grow more partial for th' observer's sake.

The Complete Poetical Works
Moral Essays
Epis. i, l. 11–12

Popper, Karl R.
Some scientists find, or so it seems, that they get their best ideas when
smoking; others by drinking coffee or whiskey. Thus there is no reason
why I should not admit that some may get their ideas by observing, or
by repeating observations.

Realism and the Aim of Science (p. 36)

Saxe, John Godfrey
It was six men of Indostan
To learning much inclined,
Who went to see the Elephant
(Though all of them were blind),
That each by observation
Might satisfy his mind.

The Poetical Works of John Godfrey Saxe
The Parable of the Blind Men and the Elephant

Swift, Jonathan
That was excellently observ'd, say I, when I read a Passage in an Author,
where his Opinion agrees with mine. When we differ, there I pronounce
him to be mistaken.

Satires and Personal Writings
Thoughts on Various Subjects

Unknown
I am an observation.
I was captured in the Field.
My conscience said 'co-operate'
My instinct said 'don't yield'.
But I yielded up my data
Now behold my sorry plight
I'm part of a statistic
Which is not a pretty sight.

The Bootstrap and the Jackknife
Oh, the tortures I've endured
They analyze my variance
Until my meaning is obscured.
But I've a plan to beat them
I'll climb up in the trees
Pretend I am a chi-square
And get freedom by degrees.

Source unknown

Whitehead, Alfred North
We habitually observe by the method of difference. Sometimes we see an elephant, and sometimes we do not. The result is that an elephant, when present, is noticed.

Process and Reality: An Essay in Cosmology (p. 6)

'Tis here, 'tis there, 'tis gone.

An Introduction to Mathematics
Chapter 1 (p. 1)

OPINION

Lippmann, Walter
True opinions can prevail only if the facts to which they refer are known; if they are not known, false ideas are just as effective as true ones, if not a little more effective.

Liberty and the News (p. 71)

Locke, John
New opinions are always suspected, and usually opposed, without any other reason but because they are not already common.

An Essay Concerning Human Understanding
Dedicatory epistle

Terence
So many men, so many opinions.

Phormio
l. 454

Twain, Mark
Our opinions do not really blossom into fruition until we have expressed them to someone else.

In Opie Read
Mark Twain and I
Five Quarts of Moonlight Juice (p. 38)

Opinions based upon theory, superstition, and ignorance are not very precious.

Letter to J.H. Twitchell
1/27/1900

Unknown
. . . opinions ought to count by weight rather than by number . . .

In James Joseph Sylvester
Collected Mathematical Works
Volume III
Address on Commemoration Day at Johns Hopkins University (p. 73)

PATENT

O'Malley, John R.
Almost every engineer is affected by the patent system.

Engineering Facts From Gatorland
Patents and the Engineer
Volume 4, Number 5, December 1967

Proverb
A patent is merely a title to a lawsuit.

In Frank Lewis Dyer
Edison: His Life and Inventions
Chapter XXVIII (p. 700)

Roosevelt, Franklin D.
Patents are the key to our technology; technology is the key to production.

In Robert A. Buckles
Ideas, Inventions, and Patents (p. 1)

PERCEPTIONS

Guest, Edgar A.
Somebody said that it couldn't be done,
  But he with a chuckle replied
That "maybe it couldn't," but he would be one
  Who wouldn't say so til he'd tried.
So he buckled right in with the trace of a grin
  On his face. If he worried he hid it.
He started to sing as he tackled the thing
  That couldn't be done, and he did it.

Somebody scoffed: "Oh, you'll never do that;
  At least no one ever has done it";
But he took off his coat and he took off his hat,
  And the first thing we knew he'd begun it.
With a lift of his chin and a bit of a grin,
  Without any doubting or quiddit.
He started to sing as he tackled the thing
  That couldn't be done, and he did it.

There are thousands to tell you it cannot be done,
  There are thousands to prophesy failure;
There are thousands to point out to you, one by one,
  The dangers that wait to assail you.
But just buckle in with a bit of a grin,
  Just take off your coat and go to it;
Just start to sing as you tackle the thing
  That "cannot be done," and you'll do it.

The Path to Home
It Couldn't Be Done

Whitehead, Alfred North
Our problem is, in fact, to fit the world to our perceptions, and not our perceptions to the world.

The Organization of Thought (p. 228)

PERSPECTIVES

Salisbury, J. Kenneth
Perspective is the quality that permits an engineer to assign correct relative importance to all things within his scope . . . An engineer without perspective is a ship without a rudder.

General Electric Review
Qualities Industry Wants in Its Engineers
May 1952

POWER

Boulton, Matthew
"Ha! Boulton," said the king. "It is long since we have seen you at court. Pray, what business are you now engaged in?"

"I am engaged, your Majesty, in the production of a commodity which is the desire of kings."

"And what is that? What is that?"

"POWER, your majesty!"

In Ralph Stein
The Great Inventions (p. 24)

Morison, George S.
Fire, animal strength, and written language have in turn advanced men and nations; something like a new capacity was developed with the discovery of explosives and again in the invention of printing; but the capacity of man has always been limited to his own individual strength and that of the men and animals he could control. His capacity is no longer so limited; man has now learned *to manufacture power*, and with the manufacture of power a new epoch began.

The New Epoch
Chapter I (p. 4)

PRAYER

Adams, Douglas
Protect me from knowing what I don't need to know. Protect me from even knowing that there are things to know that I don't know. Protect me from knowing that I decided not to know about the things that I decided not to know about. Amen.

Mostly Harmless
Chapter 9 (p. 84)

Lord, lord, lord. Protect me from the consequences of the above prayer. Amen.

Mostly Harmless
Chapter 9 (p. 84)

Ayres, C.E.
I believe in atoms, molecules, and electrons, matter of heaven and earth, and electrical energy its only form. I believe in Modern Science, conceived by Copernicus, borne out by Newton, which suffered under the inquisition, was persecuted and anathematized, but rose to the right hand of civilization as a consequence of the fact that it rules the quick and the dead. I believe in the National Research Council, the communion of the scientists, the publication of discoveries, the control of nature and the progress everlasting. Amen.

Science: The False Messiah (p. 129)

Crane, Edward V.
Father help us appreciate the unity, the power, and
   the glory of Thy system throughout the universe.
Help us to understand the unity and power of Thy
   laws, in nature, in science, and in human behavior,
Help us to sense and to fulfill the responsibility which
   we have, each of us, to Thee and Thy people,
Help us to use properly the privilege which we have in
   prayer, this privilege of coming before Thee in the

temple of our minds, to ask Thy forgiveness for our
selfish sins and shortcomings, to ask Thine aid in
cleansing and rebuilding ourselves physically, mentally,
and spiritually, and to ask Thy guidance and
Thine inspiration each day as we undertake our
chosen services and seek attunement with Thy will
and Thy wisdom.

American Engineer
Volume 26, Number 8, August 1956 (p. 47)

Florman, Samuel C.
Dear Lord, I know that I am unworthy, I confess that I have sinned,
but why did you have to abandon me on this island with nobody for
company but these boring engineers?

The Civilized Engineer (p. 10)

Hill, A.V.
Blessed are they who remain innocently in their laboratories and
grumble: for it is a thankless task to try to put things right.

The Ethical Dilemma of Science and Other Writings (p. 55)

Huxley, Thomas H.
God give me the strength to face a fact though it slay me.

In George Seldes
The Great Quotations (p. 344)

Lederman, Leon
Dear Lord, forgive me the sin of arrogance, and Lord, by arrogance I
mean the following . . .

In John D. Barrow
The Artful Universe (p. 31)

Lewis, Sinclair
God give me the unclouded eyes and freedom from haste. God give
me quiet and relentless anger against all pretense and all pretentious
work and all work left slack and unfinished. God give me a restlessness
whereby I may neither sleep nor accept praise till my observed results
equal my calculated results or in pious glee I discover and assault my
error. God give me the strength not to trust to God.

Arrowsmith
Chapter XXVI, section II (p. 292)

Poincaré, Henri
Why is it that showers and even storms seem to come by chance, so
that many people think it quite natural to pray for rain or fine weather,
though they would consider it ridiculous to ask for an eclipse by prayer?

Science and Method

Russell, Bertrand
Almighty and most merciful Machine, we have erred and strayed from thy ways like lost screws; we have put in those nuts which we ought not to have put in, and we have left out those nuts which we ought to have put in, and there is not cogginess in us . . .

The Impact of Science on Society
Chapter IV (p. 77)

Tukey, John W.
The physical sciences are used to "praying over" their data, examining the same data from a variety of points of view. This process has been very rewarding, and has led to many extremely valuable insights. Without this sort of flexibility, progress in physical science would have been much slower. Flexibility in analysis is often to be had honestly at the price of a willingness not to demand that what has *already* been observed shall establish, or prove, what analysis *suggests*. In physical science generally, the results of praying over the data are thought of as something to be put to further test in another experiment, as indications rather than conclusions.

The Annals of Mathematical Statistics
The Future of Data Analysis (p. 46)
Volume 33, Number 1, March 1962

Unknown
Grant, oh God, Thy benedictions
On my theory's predictions
Lest the facts, when verified,
Show Thy servant to have lied.

Proceedings of the Chemical Society
January 1963 (pp. 8–10)

God grant that no one else has done
    The work I want to do,
Then give me the wit to write it up
    In decent English, too.

Applied Optics
Of Optics and Opticists (p. 273)
Volume 8, Number 2, February 1969

PRECISION

Queneau, Raymond
In a bus of the S-line, 10 meters long, 3 wide, 6 high, at 3 km 600 m from its starting point, loaded with 48 people, at 12:17 p.m., a person of the masculine sex aged 27 years 3 months and 8 days, 1 m 72 cm tall and weighing 65 kg and wearing a hat 35 cm in height round the crown of which a ribbon 60 cm long, interpellated a man aged 48 years 4 months and 3 days, 1 m 68 cm tall and weighing 77 kg, by means of 14 words whose enunciation lasted 5 seconds and which alluded to some involuntary displacements of from 15 to 20 mm. Then he went and sat down about 1 m 10 cm away. 57 minutes later he was 10 meters away from the suburban entrance to the Gare Saint-Lazare and was walking up and down over a distance of 30 m with a friend aged 28, 1 m 70 cm tall and weighing 71 kg who advised him in 15 words to move by 5 cm in the direction of the zenith a button which was 3 cm in diameter.

Exercises in Style
Precision (p. 37)

Thompson, Sir D'Arcy Wentworth
Numerical precision is the very soul of science.

On Growth and Form
Chapter 1

PREDICT

Hacking, Ian

Cutting up fowl to predict the future is, if done honestly and with as little interpretation as possible, a kind of randomization. But chicken guts are hard to read and invite flights of fancy or corruption.

The Emergence of Probability (p. 3)

Kaplan, Abraham

. . . if we can predict successfully on the basis of a certain explanation we have good reason, and perhaps the best sort of reason, for accepting the explanation.

The Conduct of Inquiry
Chapter IX, section 40 (p. 350)

Unknown

To predict is one thing. To predict correctly is another.

Source unknown

PROBLEM

Alger, John R.M.
Hays, Carl V.
. . . a problem in the stage of being "recognized" is a highly emotional subject.

Creative Synthesis in Design (p. 13)

Anderson, Poul
I have yet to see any problem, however complicated, which, when you looked at it in the right way, did not become still more complicated.

New Scientist
25 September 1969 (p. 638)

Berkeley, Edmund C.
Most problems have either many answers or no answer. Only a few problems have a single answer.

Computers and Automation
Right Answers—A Short Guide for Obtaining Them
September 1969

Bloch, Arthur
Inside every large problem is a small problem struggling to get out.

Murphy's Law
Hoare's Law of Large Problems (p. 50)

Bragg, William L.
I am sure that when the first circumnavigators of the world returned from their voyage they were told by friends that some Greek philosopher . . . had held that the world was round and that they might have spared their trouble. The world is either round or flat, and endless discussion might have been carried on for ages between opposing schools who held one view or the other. The real contribution to settling the problem was made by the circumnavigators.

Science
The Physical Sciences (p. 237)
16 March 1934

Chesterton, Gilbert Keith
It isn't that they can't see the solution. It is that they can't see the problem.

The Scandal of Father Brown
The Point of the Pin

Cross, Hardy
In general the problems of civil engineers are given to them by God Almighty. They are the problems of nature. On the other hand mechanical and electrical work has problems which man, to a certain extent, has created for himself.

In Lenox H. Lohr
Centennial of Engineering 1852–1952
Professional Aspects of Mechanical Engineering (p. 150)

Easton, Elmer C.
All of the problems with which engineers are normally concerned have to do with the satisfying of some human want.

Ceramic Age
An Engineering Approach to Creative Thinking
September 1955 (p. 28)

Ehrenberg, A.S.C.
Many problems arise year after year. The answers, if only we knew them, should therefore also be similar year after year.

Data Reduction (p. 56)

Feynman, Richard P.
If we want to solve a problem that we have never solved before, we must leave the door to the unknown ajar.

What Do You Care What Other People Think (p. 2)

Frazier, A.W.
Often problems not solved earlier have not been posed earlier.

Hydrocarbon Processing
The Practical Side of Creativity
Volume 45, Number 1, January 1966

Fredrickson, A.G.
To be aware that a problem exists is the prerequisite for any attempt to solve the problem.

Chemical Engineering Education
The Dilemma of Innovating Societies (p. 148)
Volume 4, Summer 1969

HAL 9000
Sorry to interrupt the festivities, but we have a problem.

In Arthur C. Clarke
2001
Abyss, Chapter XXI (p. 73)

Herstein, I.N.
The value of a problem is not so much in coming up with the answer as in the ideas and attempted ideas it forces on the would-be solver.

Topics in Algebra (p. viii)

Hilbert, David
As long as a branch of science offers an abundance of problems, so long it is alive; a lack of problems foreshadows extinction or the cessation of independent development.

Bulletin American Mathematical Society
Hilbert: Mathematical Problems
Volume 8 (p. 438)

Hodnett, Edward
. . . being able to predict which problems you are not likely to solve is good for your peace of mind.

The Art of Problem Solving (p. 6)

You have to identify the *real* problem, and you have to identify the *total* problem.

The Art of Problem Solving (p. 12)

An unstated problem cannot be solved. Many problems go unsolved for centuries for lack of adequate statement.

The Art of Problem Solving (p. 19)

Problems often boil down to the simple form of a dilemma. A dilemma presents a choice of two solutions to a problem, both of which are unsatisfactory.

The Art of Problem Solving (p. 63)

Holmes, Sherlock
Every problem becomes very childish when once it is explained to you
. . .

In Arthur Conan Doyle
The Complete Sherlock Holmes
The Adventure of the Dancing Men

It is quite a three-pipe problem.

In Arthur Conan Doyle
The Complete Sherlock Holmes
The Red-Headed League

Juster, Norton
"I'm not very good at problems," admitted Milo.

"What a shame," sighed the Dodecahedron. "They're so very useful. Why, did you know that if a beaver two feet long with a tail a foot and a half long can build a dam twelve feet high and six feet wide in two days, all you would need to build the Kariba Dam is a beaver sixty-eight feet long with a fifty-one-foot tail?"

"Where would you find a beaver as big as that?" grumbled the Humbug as his pencil point snapped.

"I'm sure I don't know," he replied, "but if you did, you'd certainly know what to do with him."

"That's absurd," objected Milo, whose head was spinning from all the numbers and questions.

"That may be true," he acknowledged, "but it's completely accurate, and as long as the answer is right, who cares if the question is wrong? If you want sense, you'll have to make it yourself."

The Phantom Tollbooth (pp. 174–5)

Kettering, Charles Franklin
But in picking that problem be sure to analyze it carefully to see that it is worth the effort. It takes just as much effort to solve a useless problem as a useful one.

Short Stories of Science and Invention (p. 11)

I often think we have so many facilities that we lose track of the problem. Problems, as you know, are solved in the mind of some intensely interested person.

Short Stories of Science and Invention
Christman Lecturer (p. 57)

No one should pick a problem, or make a resolution, unless he realizes that the *ultimate value* of it will offset the inevitable discomfort and trouble that always goes along with the accomplishment of anything worth while. So let us not waste our time and effort on some trivial thing.

Short Stories of Science and Invention
Patience (p. 59)

A problem is not solved in a laboratory. It is solved in some fellow's head. All the apparatus is for is to get his head turned around so that he can see the thing right.

In T.A. Boyd
Professional Amateur (pp. 102–3)

Pólya, George
Only such problems come back improved whose solution we passionately desire, or for which we have worked with great tension; conscious effort and tension seem to be necessary to set the subconscious work going. At any rate, it would be too easy if it were not so; we should solve difficult problems just by sleeping and waiting for a bright idea.

Past ages regarded a sudden good idea as an inspiration, a gift of the gods. You must deserve such a gift by work, or at least by a fervent wish.

How to Solve It (p. 172)

Solving problems is a practical art, like swimming, or skiing, or playing a piano; you can learn it only by imitation and practice . . .

Mathematical Discovery
Volume I
Preface (p. v)

Popper, Karl R.
All this means that a young scientist who hopes to make discoveries is badly advised if his teacher tells him: 'Go round and observe' and that he is well advised if his teacher tell him: 'Try to learn what people are discussing nowadays in science. Find out where difficulties arise, and take an interest in disagreements. These are the questions which you should take up.' In other words, you should study the problems of the day. This means that you pick up, and try to continue, a line of inquiry which has the whole background of the earlier development of science behind it.

Conjectures and Refutations
Chapter 4 (p. 129)

Rabinow, Jacob
. . . the creating of the problem is as big an invention as the solving of the problem—sometimes, a much greater invention.

In Daniel V. DeSimone
Education for Innovation
The Process of Invention (p. 84)

Roszak, Theodore
If a problem does not have a technical solution, it must not be a *real* problem. It is but an illusion . . . a figment born of some regressive cultural tendency.

The Making of a Counter Culture
Chapter I (p. 10)

Shaw, George Bernard
All problems are finally scientific problems.

The Doctor's Dilemma
Preface
The Technical Problem

Simon, Herbert
The capacity of the human mind for formulating and solving complex problems is very small compared with the size of problems whose solution is required for objectively rational behavior in the real world—or even for a reasonable approximation to such objective rationality.

Models of Man: Social and Rational
Part IV (p. 198)

Simpson, N.F.
And suppose we solve all the problems it presents? What happens? We end up with more problems than we started with. Because that's the way problems propagate their species. A problem left to itself dries up or goes rotten. But fertilize a problem with a solution—you'll hatch out dozens.

New English Dramatists 2
A Resounding Tinkle
Act I, scene 1 (pp. 80–1)

Unknown
The situation is complicated and its difficulties are enhanced by the impossibility of saying everything at once.

Source unknown

Wiesner, Jerome Bert
Some problems are just too complicated for rational logical solutions. They admit of insights, not answers.

In D. Lang
New Yorker
Profiles: A Scientist's Advice II
26 January 1963

PROJECT

Derleth, Charles Jr.
Great public projects in their pioneering days frequently meet with criticism and obstruction, due to conflict between proposed improvements and the existing order.

In Allen Brown
Golden Gate (p. 46)

Swift, Jonathan
. . . he had been eight years upon a project for extracting sun-beams out of cucumbers; which were to be put into vials, hermetically sealed, and let out to warm the air, in raw inclement summers.

Gullivers's Travels
Part III, Chapter V

Woodson, Thomas T.
To pursue an ill-omened, unpromising project is to "continue forcing vitamins and medicine into a dead horse".

Introduction to Engineering Design (p. 57)

PROPORTION

da Vinci, Leonardo
Proportion is not only found in numbers and measurements but also in sounds, weights, times, positions, and in whatsoever power there may be.

The Notebooks of Leonardo da Vinci
Volume I
Mathematics (p. 636)

REALITY

Bohr, Niels
. . . an independent reality in the ordinary physical sense can neither be ascribed to the phenomena nor to the agencies of observation.

Atomic Theory and the Description of Nature
Development of Atomic Theory (p. 54)

Einstein, Albert
Physical concepts are free creations of the human mind, and are not, however it may seem, uniquely determined by the external world. In our endeavor to understand reality we are somewhat like a man trying to understand the mechanism of a closed watch. He sees the face and the moving hands, even hears its ticking but he has no way of opening the case. If he is ingenious he may form some picture of a mechanism which could be responsible for all the things he observes but he may never be quite sure his picture is the only one which could explain his observations. He will never be able to compare his picture with the real mechanism and he cannot even imagine the possibility or the meaning of such a comparison.

The Evolution of Physics (p. 33)

Unknown
My reality check just bounced.

Source unknown

Wheeler, John A.
What we call reality consists . . . of a few iron posts of observation between which we fill an elaborate papier-mâché of imagination and theory.

In Harry Woolf (Editor)
Some Strangeness in the Proportion
Beyond the Black Hole
Chapter 22

REASON

Beveridge, W.I.B.

How easy it is for unverified assumptions to creep into our reasoning unnoticed!

The Art of Scientific Investigation (p. 87)

Drummond, Sir William

. . . he, who will not reason, is a bigot; he, who cannot, is a fool; and he, who dares not, is a slave.

Academical Questions
Preface (p. xv)

Johnson, Samuel

We may take Fancy for a companion, but must follow Reason as our guide.

Letter to Boswell 1774

Romanoff, Alexis L.

Reasoning goes beyond the analysis of facts.

Encyclopedia of Thoughts
Aphorisms
1973

Shakespeare, William

His reasons are as two grains of wheat hid in two bushels of chaff: you shall seek all day ere you find them, and when you have them, they are not worth the search.

The Complete Works of William Shakespeare
The Merchant of Venice
Act I, scene 1, l. 115

Good reason must, of force, give place to better.

Julius Caesar
Act IV, scene 3, l. 203

Whitehead, Alfred North

The art of reasoning consists in getting hold of the subject at the right end, of seizing on the few general ideas that illuminate the whole, and of persistently organizing all subsidiary facts round them. Nobody can be a good reasoner unless by constant practice he has realized the importance of getting hold of the big ideas and hanging on to them like grim death.

Presidential Address to the London Branch of
the Mathematical Association, 1914
In W.W. Sawyer
Prelude to Mathematics (p. 183)

When all else fails
Use bloody great nails.
Unknown – (See p. 240)

REPAIR

Adams, Douglas
The major difference between a thing that might go wrong and a thing that cannot possibly go wrong is that when a thing that cannot possibly go wrong goes wrong, it usually turns out to be impossible to get at and repair.

Mostly Harmless
Chapter 12 (pp. 137–8)

REPORTS

Kettering, Charles Franklin
Some technical reports are so dry and dusty . . . that if you put a pile of them in a hydraulic press and apply millions of pounds of pressure to it, not a drop of juice will run out.

Professional Amateur (p. 215)

Smith, Ralph J.
The word 'report' means to 'carry back', to bring back information about something seen or investigated.

Engineering as a Career (p. 212)

RESEARCH

Aurelius, Marcus [Antoninus]
Nothing has the power to broaden the mind as the ability to investigate systematically and truly all that comes under thy observations in life.

The Meditations of the Emperor Antoninus Marcus Aurelius
Book iii, section 2

Bates, Marston
Research is the process of going up alleys to see if they are blind.

In Jefferson Hane Weaver
The World of Physics
Volume II (p. 63)

Bradley, A.C.
Research, though toilsome, is easy; imagination, though delightful, is difficult.

Oxford Lectures on Poetry

Bush, V.
Basic research leads to new knowledge. It provides scientific capital. It creates the fund from which the practical applications of knowledge must be drawn. New products and new processes do not appear full-grown. They are founded on new principles and new conceptions, which in turn are painstakingly developed by research in the purest realms of science.

Endless Horizons (pp. 52–3)

Chesterton, Gilbert Keith
Research is the search of people who don't know what they want.

The G.K. Chesterton Calendar
May Twenty-fifth

da Vinci, Leonardo
Nothing is written as the result of new researches.

Leonardo da Vinci's Notebooks (p. 53)

Green, Celia

The way to do research is to attack the facts at the point of greatest astonishment.

The Decline and Fall of Science
Aphorisms (p. 1)

Gregg, Alan

One wonders whether the rare ability to be completely attentive to, and to profit by, Nature's slightest deviation from the conduct expected of her is not the secret of the best research minds and one that explains why some men turn to most remarkably good advantage seemingly trivial accidents. Behind such attention lies an unremitting sensitivity.

The Furtherance of Medical Research

Kettering, Charles Franklin

We find that in research a certain amount of intelligent ignorance is essential to progress; for if you know too much, you won't try the thing.

In T.A. Boyd
Professional Amateur (p. 106)

Matthew 7:7

Seek and ye shall find; knock, and it shall be opened unto you.

The Bible

RESEARCH PLANS

Richter, Curt P.
Good researchers use research plans merely as starters and are ready to scrap them at once in the light of actual findings.

Science
Volume 118, 1953 (p. 92)

van Noordwijk, J.
. . . however excellent multiannual planning, research-project management, and time recording may be, the scientist should always have some opportunity to test the idea that he got that morning while shaving.

Perspectives in Biology and Medicine
The Bioassayist (p. 307)
Volume 29, Number 2, Winter 1986

Waksman, Selman A.
. . . a new problem has arisen—namely 'planned research' versus the 'individual investigator'. There is place for planned research. It can take a defined body of knowledge and lay out a set of experiments which will exploit this knowledge to its foreseeable limits. It can take a set of postulates and drive them home to their logical conclusions. It can do this with exhaustive thoroughness, economy, and speed. Within its limitations, it is efficient, expeditious, and authoritative. But there is a place also and a more important place for the random investigator. The role of planned research is to consolidate ground already won; the role of the random investigator is to seek out new worlds to conquer.

Perspectives in Biology and Medicine
Volume 7, 1964 (p. 311)

Good scientists use research plans merely as outlines to begin their investigations and are ready to give them up once they are not justified by actual findings. Experimental designs tend to give rise to 'team research', which serves a purpose in developing and applying ideas; it rarely produces new ideas.

Perspectives in Biology and Medicine
Volume 7, 1964 (p. 312)

HER FUTURE'S LOOKING DIM..!

Cutting up fowl to predict the future is, if done honestly and with as little interpretation as possible, a kind of randomization. But chicken guts are hard to read and invite flights of fancy or corruption.
Ian Hacking – (See p. 213)

RISK

Fisher, Irving
Risk varies inversely with knowledge.

The Theory of Interest
Chapter IX (p. 221)

Florman, Samuel C.
Good intentions and high moral standards do not help an engineer establish the limits of acceptable risk.

Blaming Technology
Moral Blueprints (p. 173)

RULE

Burton, Robert
No rule is so generall, which admits not some exception . . .

Anatomy of Melancholy
Part I, section ii, mem. 2, subs. 3 (p. 225)

Cervantes, Miguel de
There is no rule without an exception.

Don Quixote
Part ii, Chapter 18

Edison, Thomas
Hell! there ain't no rules around here! We are tryin' to accomplish somep'n!

In Robert Byrne
The Fourth, and By Far the Most Recent, 637 Best Things Anybody Ever Said
Famous American Men of Science

Galsworthy, John
KEITH: . . . I don't see the use in drawin' hard and fast rules. You only have to break 'em.

Eldest Son
Act I, scene 2 (p. 13)

Wilson, John
. . . the Exception proves the Rule.

The Cheats
Appendix
The Author to the Reader
l. 27

SEEING

Marsh, George Perkins
Sight is a faculty; seeing, an art.

Man and Nature
Introductory (p. 15)

Wilde, Oscar
To look at a thing is very different from seeing a thing.

Phrases and Philosophies for the use of the Young
Sebastian Melmoth (p. 123)

SIMPLICITY

Wright, Frank Lloyd
To know what to leave out and what to put in; just where and just how, ah, *that* is to have been educated in knowledge of simplicity . . .

Frank Lloyd Wright: An Autobiography
Simplicity (p. 144)

See Saw, Marjorie Daw,
She rocked—and learned the lever law.
She saw that he weighed more than she
For she sat higher up than he,...
L. A. Graham – (See p. 175)

SOLIDITY

Locke, John
The idea of solidity we receive by our touch: and it arises from the resistance which we find in a body to the entrance of any other body into the place it possesses, till it has left it. There is no idea which we receive more constantly from sensation than solidity. Whether we move or rest, in what posture soever we are, we always feel something under us that supports us, and hinders our further sinking downwards; and the bodies which we daily handle make us perceive that, whilst they remain between them, they do, by an insurmountable force, hinder the approach of the parts of our hands that press them. That which thus hinders the approach of two bodies, when they are moved one towards another, I call solidity.

An Essay Concerning Human Understanding
Book II, Chapter 4, section 1

SOLUTION

Arber, Agnes
. . . a problem put aside in one period must have the right interval of dormancy before awakening, freshened, to an unforced solution, when the time is naturally ripe.

The Mind and the Eye
Chapter I (p. 7)

Asimow, Morris
A *solution* is a synthesis of component elements which hurdles the obstructing difficulties and, neither exceeding the available resources nor encroaching on the limits set by the constraints, accomplishes the prescribed goals.

Introduction to Engineering Design (p. 45)

Hodnett, Edward
The solution that is not ready in time is not a solution.

The Art of Problem Solving (p. 80)

King, Blake
Recognize that every work of man represents an imperfect solution to the problem.

Mechanical Engineering
Object: Creativity (p. 41)
November 1963

Wellington, Arthur Mellen
As the correct solution of any problem depends primarily on a true understanding of what the problem really is, and wherein lies its difficulty, we may profitably pause upon the threshold of our subject to consider first, in a more general way, its real nature; the causes which impede sound practice; the conditions on which success or failure depends; the directions in which error is most to be feared.

The Economic Theory of Railway Location (p. 1)

SPECIALIZATION

Compton, Karl Taylor
There are two principal reasons for increasing training in scientific principles and the more fundamental aspects of engineering, as opposed to going farther into the field of specialization in the training of engineers. The first of these reasons is to be found in the development of the engineering specialties themselves, starting first with simply applied science followed next by the branching off into the various engineering fields, such as electrical, mechanical, civil, etc, until with further discoveries and developments there arose subdivisions of these fields, such as radio, hydraulic, refrigeration engineering, etc. Now even these subdivisions are becoming so highly specialized as to indicate the need of still further subdivision in the direction of specialization.

A Scientist Speaks (p. 51)

SPECIFICATIONS

Alger, John R.M.
Hays, Carl V.
Once a problem is recognized clearly and all the parties concerned have agreed on its nature, the development of detailed specifications becomes vital.

Creative Synthesis in Design (p. 15)

A good engineer is adroit in negotiating changes in specifications or trade-offs.

Creative Synthesis in Design (p. 16)

Hoover, Herbert
Specifications are the formulated, definite, and complete statements of what the buyer requires of the seller.

National Directory of Commodity Specifications
M 65
Foreword (p. 1)

Mathews, J.A.
Good sense is highly desirable in writing specifications and is even more necessary in interpreting them. If they only contained the minimum number of requirements to define the character of material wanted . . . the matter would be greatly simplified. Rarely do they cover the only material suited to the purpose intended, and more rarely do they cover the best material for the purpose intended. Once written, they become as the laws of the Medes and Persians, which alter not. They acquire a sort of sanctity, like the Ten Commandments or the Constitution before the adoption of the Eighteenth Amendment.

Mechanical Engineering
Present Tendencies in Engineering Materials (p. 792)
Volume 48, Number 8, August 1926

Unknown
It is easier to change the specification to fit the product than vice versa.

Source unknown

STABILITY

Unknown

Stability is only good if you aren't heading straight into the ground.

Source unknown

STRENGTH

Faraday, Michael
Now, how curiously our ideas expand by watching these conditions of the attraction of cohesion!—how many new phenomena it gives us beyond those of the attraction of gravitation! See how it gives us great strength. The things we deal with in building up the structures on the earth are of great strength (we use iron, stone and other things of great strength); and only think that all those structures you have about you—think of the "Great Eastern", if you please, which is of such size and power as to be almost more than man can manage—are the result of this power of cohesion and attraction.

On the Various Forces of Nature
Lecture II (p. 56)

Unknown
When all else fails
Use bloody great nails.

In James E. Gordon
The New Science of Strong Materials (p. 144)

SURVEYOR

Galton, Francis

I thought it safer to proceed like the surveyor of a new country, and endeavour to fix in the first instance as truly as I could the position of several cardinal points.

In Carl Pearson
The Life, Letters and Labours of Francis Galton
Volume II (p. xii)

Kafka, Franz

You've been taken on as Land Surveyor, as you say, but, unfortunately, we have no need of a Land Surveyor. There wouldn't be the least use for one here. The frontiers of our little state are marked out and all officially recorded.

The Castle
Chapter 5 (p. 77)

SYMMETRY

Blake, William
Tyger, Tyger, burning bright
In the forest of the night,
What immortal hand or eye
Could frame thy fearful symmetry?

The Complete Writings of William Blake
Songs of Innocence and of Experience
The Tyger

Borges, Jorge Luis
Reality favors symmetry.

Conversations with Jorge Luis Borges
Edited by Richard Durgin

Feynman, Richard P.
Why is nature so nearly symmetrical? No one has any idea why. The only thing we might suggest is something like this: There is a gate in Japan, a gate in Neiko, which is sometimes called by the Japanese the most beautiful gate in all Japan; it was built in a time when there was great influence from Chinese art. The gate is very elaborate, with lots of gables and beautiful carvings and lots of columns and dragon heads and princes carved into the pillars, and so on. But when one looks closely he sees that in the elaborate and complex design along one of the pillars, one of the small design elements is carved upside down; otherwise the thing is completely symmetrical. If one asks why this is, the story is that it was carved upside down so that the gods will not be jealous of the perfection of man. So they purposely put the error in there, so that the gods would not be jealous and get angry with human beings.

We might like to turn the idea around and think that the true explanation of the near symmetry of nature is this: that God made the laws only nearly symmetrical so that we should not be jealous of His perfection!

The Feynman Lectures on Physics
Volume I, 52-9 (p. 52-12)

Mao Tse-tung
Tell me why should symmetry be of importance?

To Nobel Laureate Tsung-Dao Lee
30 May 1974

Updike, John

When you look	kool uoy nehW
into a mirror	rorrim a otni
it is not	ton si ti
yourself you see,	,ees uoy flesruoy
but a kind	dnik a tub
of apish error	rorre hsipa fo
posed in fearful	lufraef ni desop
symmetry	yrtemmys

Telephone Poles and Other Poems
Mirror

Weyl, Hermann
Symmetry, as wide or as narrow as you may define its meaning, is one idea by which man through the ages has tried to comprehend and create order, beauty, and perfection.

Symmetry (p. 5)

SYSTEM

Gall, John
A Large System, Produced By Expanding The Dimensions Of A Smaller System, Does Not Behave Like The Smaller System.

Systemantics (p. 45)

A Complex System Designed From Scratch Never Works And Cannot Be Patched Up To Make It Work. You Have To Start Over, Beginning With A Simple Working System.

Systemantics (pp. 80–1)

Any Large System Is Going To Be Operating Most Of The Time In Failure Mode.

Systemantics (p. 91)

Complicated Systems Produce Complicated Responses (Not Solutions) To Problems.

Systemantics (p. 109)

Peacock, Thomas Love
All philosophers who find
Some favorite system to their mind,
In every point to make it fit
Will force all nature to submit.

Headlong Hall (p. 44)

TABLES

Benzecri, J.P.
That one sees these long lists of little tables coming out of the printers, is because too many scientists, especially in the social sciences, have not adapted their methods to the power of these new computing tools. They are like an engineer who builds a bridge by designing blocks of concrete in the form of bricks.

Lois de Probabilité sur un Ensemble Produit (p. 1)

Carlyle, Thomas
Tables are like cobwebs, like the sieve of the Danaides; beautifully reticulated, orderly to look upon, but which will hold no conclusion. Tables are abstractions . . . There are innumerable circumstances; and one circumstance left out may be the vital one on which all turned . . . Conclusive facts are inseparable from inconclusive except by a head that already understands and knows.

English and other Critical Essays
Chartism
Chapter II

Playfair, William
Information that is imperfectly acquired, is generally as imperfectly retained; and a man who has carefully investigated a printed table, finds, when done, that he has only a very faint and partial idea of what he has read; and that like a figure imprinted on sand, is soon totally erased and defaced.

The Commercial and Political Atlas (p. 3)

TECHNOLOGICAL

Hanham, H.J.
Great technological advances are always around the corner.

Daedalus
Clio's Weapons (p. 509)
Spring 1971

Harrington, Michael
. . . if there is technological advance without social advance, there is, almost automatically, an increase in human misery, in impoverishment.

The Other America
Appendix, section 1 (p. 188)

Hoffer, Eric
Where there is the necessary technical skill to move mountains, there is no need for the faith that moves mountains.

The Passionate State of Mind
Number 12 (p. 7)

Orwell, George
Men are only so good as their technical developments allows them to be.

Inside the Whale and Other Essays
Charles Dickens

TECHNOLOGY

Adler, Alfred
The confusion of science with technology is understandable. Certainly the two often appear to be aspects of a single larger process, as when science proposes new laws of physics, which inspire the development of a technology for their exploration, which in turn exposes inaccuracies in the laws and forces science to seek a more profound level of theory. But in fact their divergence is great. It is in the divergence of engagement from fulfillment, of means from ends . . . If truth is a path, then science explores it, and the brief stops along the way are where technologies begin (they build towns and pave a highway). Technology is results, science is process; though the two fuse and separate and then fuse once more, as ends and means must, their opposition is profound.

Atlantic Monthly
Volume 279, Number 2, February 1997 (p. 16)

Allen, Charles M.
If the human race wants to go to hell in a basket, technology can help it get there by jet. It won't change the desire or the direction, but it can greatly speed the passage.

Speech
Wake Forest University
Winston-Salem, North Carolina
25 April 1967

Ashby, Sir Eric
The habit of apprehending a technology in its completeness: this is the essence of technological humanism, and this is what we should expect education in the higher technology to achieve. I believe it could be achieved by making specialist studies . . . the core around which are grouped liberal studies which are relevant to these specialist studies. But they must be relevant; the path to culture should be through a man's specialism, not by-passing it . . .

A student who can weave his technology into the fabric of society can claim to have a liberal education; a student who cannot weave his technology into the fabric of society cannot claim even to be a good technologist.

Technology and the Academics
Chapter 4 (pp. 84 and 85)

Association of American Colleges
. . . we have become a people unable to comprehend the technology we invent . . .

Integrity in the College Curriculum (p. 2)
February 1985

Ballard, J.G.
Science and technology multiply around us. To an increasing extent they dictate the languages in which we speak and think. Either we use those languages, or we remain mute.

Crash
Introduction (p. 7)

Barzun, Jacques
. . . something pervasive that makes the difference, not between civilized man and the savage, not between man and the animals, but between man and the robot, grows numb, ossifies and falls away like black mortified flesh when techne assails the senses and science dominates the mind.

In T. Dobzhansky
The Biology of Ultimate Concern
Chapter 5 (p. 103)

Beer, Stafford
If it works, it's out of date.

Brain of the Firm (p. v)

Bronowski, Jacob
Every civilization has been grounded on technology: what makes ours unique is that for the first time we believe that every man is entitled to all its benefits.

The American Scholar
Technology and Culture in Evolution (p. 207)
Volume 41, Number 2, Spring 1972

Bunge, M.
. . . whereas science elicits changes in order to know, technology knows in order to elicit changes.

In G. Bugliarello and D.B. Doner (Editors)
The History And Philosophy Of Technology
Chapter 15 (p. 264)

Cardwell, D.S.L.

There are, as we have seen, a number of different modes of technological innovation. Before the seventeenth century inventions (empirical or scientific) were diffused by imitation and adaptation while improvement was established by the survival of the fittest. Now, technology has become a complex but consciously directed group of social activities involving a wide range of skills, exemplified by scientific research, managerial expertise, and practical and inventive abilities. The powers of technology appear to be unlimited. If some of the dangers may be great, the potential rewards are greater still. This is not simply a matter of material benefits for, as we have seen, major changes in thought have, in the past, occurred as consequences of technological advances.

Dictionary of the History of Ideas
Volume IV
Technology (p. 364)

Clarke, Arthur C.

Any sufficiently advanced technology is indistinguishable from magic.

The Lost Worlds of 2001
Chapter 34 (p. 189)

Commoner, Barry

Despite the dazzling successes of modern technology and the unprecedented power of modern military systems, they suffer from a common and catastrophic fault. While providing us with a bountiful supply of food, with great industrial plants, with high-speed transportation, and with military weapons of unprecedented power, they threaten our very survival.

Science and Survival (p. 126)

DeSimone, Daniel V.

Technological invention and innovation are the business of engineering. They are embodied in engineering change.

Education for Innovation
Introduction (p. 4)

Drexler, K. Eric

The promise of technology lures us onward, and the pressure of competition makes stopping virtually impossible. As the technology race quickens, new developments sweep toward us faster, and a fatal mistake grows more likely. We need to strike a better balance between our foresight and our rate of advance. We cannot do much so slow the growth of our technology, but we can speed growth of foresight. And with better foresight, we will have a better chance to steer the technology race in safe directions.

Engines of Creation
Finding the Facts (p. 203)

Dubos, René
The idealistic and the demonic forces in nationalism are as powerful today as they were in the past but their expressions are changing, because human history is moving from its hallowed parochial traditions to the era of global technology . . . The one credo of technology which has been accepted practically all over the world is that nature is to be regarded as a source of raw materials to be exploited for human ends rather than as an entity to be appreciated for its own value.

A God Within (p. 204)

Dyson, Freeman J.
If we had a reliable way to label our toys good and bad, it would be easy to regulate technology wisely. But we can rarely see far enough ahead to know which road leads to damnation. Whoever concerns himself with big technology, either to push it forward or to stop it, is gambling in human lives.

Disturbing the Universe
Chapter 1 (p. 7)

Editorial, New York Times
Technology, when misused, poisons air, soil, water and lives. But a world without technology would be prey to something worse: the impersonal ruthlessness of the natural order, in which the health of a species depends on relentless sacrifice of the weak.

New York Times
29 August 1986

Esar, Evan
Technology enables man to gain control over everything except technology.

20,000 Quips and Quotes

Technology improves things so fast that by the time we can afford the best, there's something better.

20,000 Quips and Quotes

Technology is rapidly filling our homes with appliances smarter than we are.

20,000 Quips and Quotes

Technology has made improvements in everything, except the weather and people.

20,000 Quips and Quotes

Feynman, Richard
For a successful technology, reality must take precedence over public relations, for Nature cannot be fooled.

What Do You Care What Other People Think?
Appendix F (p. 237)

Frisch, Max
Technology is the knack of so arranging the world that we don't have to experience it.

In Rollo May
The Cry for Myth (p. 57)

Gabor, Dennis
The most important and urgent problems of the technology of today are no longer the satisfactions of the primary needs or of archetypal wishes, but the reparation of the evils and damages by technology of yesterday.

Innovations: Scientific, Technological and Social (p. 9)

Galbraith, John Kenneth
The imperatives of technology and organization, not the images of ideology, are what determine the shape of economics.

The New Industrial State
Chapter I (p. 19)

It is a commonplace of modern technology that there is a high measure of certainty that problems have solutions before there is knowledge of how they are to be solved.

The New Industrial State
Chapter II, section 4 (p. 19)

Heidegger, Martin
. . . the *essence* of technology . . . is nothing technological.

Basic Writings
The Question Concerning Technology (p. 285)

Hilbert, David
One hears a good deal now-a-days about the hostility between Science and Technology. I don't think that is true gentlemen. It almost isn't true. It can't be true. *Sie haben ja gar nichts miteinander zu tun.* They have nothing whatsoever to do with one another.

In M. Gardner
Great Essays in Science
Physics in the Contemporary World (p. 194)

Huxley, Aldous
Advances in technology do not abolish the institution of war; they merely modify its manifestations.

Science, Liberty and Peace (p. 47)

Jones, Barry
The reality is that many of the changes in science and technology are complex because of the complexity of them.

Sydney Morning Herald
Sayings of the Week
12 July 1986

Kaysen, Carl
. . . the advance of technology, like the growth of population and industry, has also been proceeding exponentially.

Foreign Affairs
Limits to Growth (p. 664)
Volume 50, Number 4, July 1972

Kranzberg, Mel
Technology is neither good nor bad, nor is it neutral.

Attributed

Krutch, Joseph Wood
Technology made large populations possible; large populations now make technology indispensable.

Human Nature and Human Condition
The Nemesis of Power (p. 145)

Lerner, Max
. . . a world technology means either a world government or world suicide.

Actions and Passions
The Imagination of H.G. Wells (p. 17)

Lilienthal, David E.
The machine that frees man's back of drudgery does not thereby make his spirit free. Technology has made us more productive, but it does not necessarily enrich our lives. Engineers can build us great dams, but only great people make a valley great. There is no technology of goodness. Men must make themselves spiritually free.

TVA: Democracy on the March (p. 218)

Lovins, Amory B.
Any demanding high technology tends to develop influential and dedicated constituencies of those who link its commercial success with both the public welfare and their own. Such sincerely held beliefs, peer pressures, and the harsh demands that the work itself places on time and energy all tend to discourage such people from acquiring a similarly thorough knowledge of alternative policies and the need to discuss them.

Foreign Affairs
Energy Strategy (p. 93)
Volume 55, Number 1, October 1976

McRobie, George
The choice of technology, whether for a rich or a poor country, is probably the most important decision to be made.

Conservation Foundation Letter
October 1976 (p. 1)

Meadows, Donella H.
Meadows, Dennis L.
Randers, Jørgen
Behrens, William W., III
Technology can relieve the symptoms of a problem without affecting the underlying causes. Faith in technology as the ultimate solution to all problems can thus divert our attention from the most fundamental problem—the problem of growth in a finite system—and prevent us from taking effective action to solve it.

The Limits to Growth
Chapter IV (p. 154)

Mumford, Lewis
. . . however far modern science and technics have fallen short of their inherent possibilities, they have taught mankind at least one lesson: Nothing is impossible.

Technics and Civilization
Chapter VIII (p. 435)

Oppenheimer, Julius Robert
The open society, the unrestricted access to knowledge, the unplanned and uninhibited association of men for its furtherance—these are what may make a vast, complex, ever growing, ever changing, ever more specialized and expert technological world, nevertheless a world of human community.

Science and the Common Understanding
Chapter 6 (p. 95)

In fact, most people—when they speak of Science as a good thing—have in mind such Technology as has altered the condition of their life.

Great Essays in Science
Physics in the Contemporary World (p. 194)

Philip, Prince, Duke of Edinburgh
Our way of life has been influenced by the way technology has developed. In future, it seems to me, we ought to try to reverse this and so develop our technology that it meets the needs of the sort of life we wish to lead.

Men, Machines and Sacred Cows

Reich, Charles A.
Technology and production can be great benefactors of man, but they are mindless instruments, and if undirected they careen along with a momentum of their own. In our country, they pulverize everything in their path: the landscape, the natural environment, history and tradition, the amenities and civilities, the privacy and spaciousness of life, beauty, and the fragile, slow-growing social structures which bind us together.

The Greening of America
Chapter 1 (p. 6)

Rickover, H.G.
. . . technology can have no legitimacy unless it inflicts no harm.

Mechanical Engineering
A Humanistic Technology (p. 45)
November 1982

Schumacher, E.F.
. . . the system of nature, of which man is a part, tends to be self-balancing, self-adjusting, self-cleansing. Not so with technology . . .

The technology of *mass production* is inherently violent, ecologically damaging, self-defeating in terms of non-renewable resources, and stultifying for the human person.

Small is Beautiful
Part II, Chapter V

Snow, C.P.
Technology . . . is a queer thing. It brings you great gifts with one hand, and it stabs you in the back with the other.

New York Times
15 March 1971

Sophocles
Wonders are many, and none is more wonderful than man; the power that crosses the white sea, driven by the stormy south-wind, making a path under surges that threaten to engulf him; . . . turning the soil with the offspring of horses, as the ploughs go to and fro from year to year . . .

And speech, and windswift thought, and all the moods that mould a state, hath he taught himself; and how to flee the arrows of the frost, when 'tis hard lodging under the clear sky, and the arrows of the rushing rain; yea, he hath resource for all . . .

Antigone
l. 333–340, 349–354

Soule, Michael

Since we have no choice but to be swept along by this vast technological surge, we might as well learn to surf.

In David Western and Mary C. Pearl
Conservation for the Twenty-first Century
Conservation Biology in the Twenty-first Century:
Summary and Outlook (p. 303)

Stevenson, Adlai E.

Technology, while adding daily to our physical ease, throws daily another loop of fine wire around our souls. It contributes hugely to our mobility, which we must not confuse with freedom. The extensions of our senses, which we find so fascinating, are not adding to the discrimination of our minds, since we need increasingly to take the reading of a needle on a dial to discover whether we think something is good or bad, or right or wrong.

Fortune Magazine
My Faith in Democratic Capitalism (p. 156)
October 1955

Toffler, Alvin

. . . that great, growling engine of change—technology.

Future Shock
The Accelerative Thrust
The Technological Engine (p. 25)

. . . technology feeds on itself. Technology makes more technology possible . . .

Future Shock
The Accelerative Thrust
The Technological Engine (p. 27)

If technology . . . is to be regarded as a great engine, a mighty accelerator, then knowledge must be regarded as its fuel. And we thus come to the crux of the accelerative process in society, for the engine is being fed a richer and richer fuel every day.

Future Shock
The Accelerative Thrust
The Technological Engine (pp. 29–30)

TESTING

da Vinci, Leonardo
It is by testing that we discern fine gold.

Leonardo da Vinci's Notebooks (p. 60)

THEOREM

Brown, Gordon Spencer

A theorem is no more proved by logic and computation than a sonnet is written by grammar and rhetoric, or than a sonata is composed by harmony and counterpoint, or a picture painted by balance and perspective. Logic and computation, grammar and rhetoric, harmony and counterpoint, balance and perspective, can be seen in the work *after* it is created, but these forms are, in the final analysis, parasitic on, they have no existence apart from, the creativity of the work itself. Thus the relation of logic to mathematics is seen to be that of an applied science to its pure ground, and all applied science is seen as drawing sustenance from a process of creation with which it can combine to give structure, but which it cannot appropriate.

Laws of Form (p. 102)

Huxley, Aldous

Too much theorizing, as we all know, is fatal to the soul . . .

Tomorrow and Tomorrow and Tomorrow
The Education of an Amphibian (p. 7)

Poincaré, Henri

I beg your pardon; I am about to use some technical expressions, but they need not frighten you for you are not obliged to understand them. I shall say, for example, that I have found the demonstration of such a theorem under such circumstances. This theorem will have a barbarous name unfamiliar to many, but that is unimportant; what is of interest for the psychologist is not the theorem but the circumstances . . .

The Foundations of Science
Mathematical Creation

THERMODYNAMICS

Cardenal, Ernesto
The second law of thermodynamics!:
energy is indestructible in quantity
but continually changes in form.
    And it always runs down like water.

Cosmic Canticle
Cantiga 3
Autumn Fugue

Epstein, P.S.
Thermodynamics deals with systems which, in addition to mechanical
and electromagnetic parameters, are described by a specifically thermal
one, namely, the temperature or some equivalent of it. Thermodynamics
is essentially a science about the conditions of equilibrium of systems
and about the processes which can go on in states little different from
the state of equilibrium.

Textbook of Thermodynamics
Chapter I (p. 2)

Meixner, J.
A careful study of the thermodynamics of electrical networks has given
considerable insight into these problems and also produced a very
interesting result: the non-existence of a unique entropy value in a state
which is obtained during an irreversible process . . . I would say, I have
done away with entropy. The next step might be to let us also do away
with temperature.

In Edward B. Stuart, Benjamin Gal-Or and Alan J. Brainard (Editors)
A Critical Review of Thermodynamics

Stenger, Victor J.
Scientists speak of the Law of Inertia or the Second Law of
Thermodynamics as if some great legislature in the sky once met and
set down rules to govern the universe.

Not By Design (p. 14)

TIME

Bergson, Henri
Time is invention or it is nothing at all.

Creative Evolution (p. 341)

Bondi, Hermann
Time must never be thought of as pre-existing in any sense; it is a manufactured quantity.

In Paul Davies
About Time (p. 22)

Clemence, G.M.
The measurement of time is essentially a process of counting.

The American Scientist
Time and its Measurement (p. 261)
Volume 40, Number 2, April 1952

Fraser, J.T.
The resulting dichotomy between time felt and time understood is a hallmark of scientific–industrial civilization, a sort of collective schizophrenia.

In Eric J. Lerner
The Big Bang Never Happened (p. 283)

The Rocky Horror Picture Show
With a bit of a mind flip
You're into the time slip
And nothing can ever be the same.

The Time Warp

Saint Augustine
Time is like a river made up of events which happen, and its current is strong; no sooner does anything appear than it is swept away.

In Paul Davies
Other Worlds (p. 186)

Unknown
Time is just one damn thing after another.

Source unknown

Time is God's way of keeping things from happening all at once.

Source unknown

Technology is rapidly filling our homes with appliances smarter than we are.
Evan Esar – (See p. 250)

TOOL

Beecher, Henry Ward
A tool is but the extension of a man's hand, and a machine is but a complex tool. He that invents a machine augments the power of a man and the well-being of mankind.

In Lenox R. Lohr
Centennial of Engineering 1852–1952
Historical Background (p. 340)

Bergson, Henri
Science has equipped man in less than fifty years with more tools than he had made during the thousands of years he had lived on earth. Each new machine being for man a new organ—an artificial organ—his body became suddenly and prodigiously increased in size, without his soul being at the same time able to dilate to the dimensions of his body.

In Lenox R. Lohr
Centennial of Engineering 1852–1952
Historical Background (p. 343)

Carlyle, Thomas
Man is a Tool-using Animal (*Handthierendes Thier*). Weak in himself, and of small stature, he stands on a basis, at most for the flattest-soled, of some half square foot, insecurely enough; Has to straddle out his legs, lest the very wind supplant him. Feeblest of bipeds! Three quintals are a crushing load for him; the steer of the meadow tosses him aloft like a waste rag. Nevertheless he can use Tools: with these the granite mountain melts into light dust before him, seas are his smooth highway, winds and fire his unwearying steeds. Nowhere do you find him without Tools; without Tools he is nothing, with Tools he is all.

Sartor Resartus
Chapter 5 (p. 34)

TRAIN ENGINEER

Dalhart, Vernon
The engineer pulled at the whistle
For the brakes wouldn't work when applied
And the brakeman climbed out on the car tops
For he knew what the whistle had cried.

The Altoona Freight Wreck
Written by Fred Tait-Douglas and Carson Robison
From *Scalded to Death by the Steam*

Unknown
Come all you rounders that want to hear
The story of a brave engineer.
Casey Jones was the rounder's name,
On a six eight wheeler, boys, he won his fame.

http://www.mudcat.org
Casey Jones

With hand upon the lever and eye upon the track
The engineer is standing while the shades of night grow black

http://www.mudcat.org
The Chatsworth Wreck

TRUTH

Bacon, Francis
Truth emerges more readily from error than from confusion.

Novum Organum
In Ritchie Calder
Man and the Cosmos (p. 19)

Bohr, Niels
The opposite of a correct statement is a false statement. But the opposite of a profound truth may well be another profound truth.

In Werner Heisenberg
Physics and Beyond (p. 102)

Bronowski, Jacob
We cannot define truth in science until we move from fact to law. And within the body of laws in turn, what impresses us as truth is the orderly coherence of the pieces. They fit together like the characters of a great novel, or like the words of a poem. Indeed, we should keep that last analogy by us always, for science is a language, and like a language it defines its parts by the way they make up a meaning. Every word in a sentence has some uncertainty of definition, and yet the sentence defines its own meaning and that of its words conclusively. It is the internal unity and coherence of science which gives it truth, and which makes it a better system of prediction than any less orderly language.

The Common Sense of Science
Truth and Value (p. 131)

Heaviside, Oliver
We do not dwell in the Palace of Truth. But, as was mentioned to me not long since, "There is a time coming when all things shall be found out." I am not so sanguine myself, believing that the well in which Truth is said to reside is really a bottomless pit.

Electromagnetic Theory
Chapter I, Volume I (p. 1)

Heinlein, Robert A.

The hardest part about gaining any new idea is sweeping out the false idea occupying that niche. As long as that niche is occupied, evidence and proof and logical demonstration get nowhere. But once the niche is emptied of the wrong idea that has been filling it—once you can honestly say, 'I don't know,' then it becomes possible to get at the truth.

The Cat Who Walks Through Walls (p. 244)

Lawson, Alfred William

Education is the science of knowing TRUTH.
Miseducation is the art of absorbing FALSITY.
TRUTH is that which is, not that which ain't.
FALSITY is that which ain't, not that which is.

In Martin Gardner
Fads and Fallacies (p. 76)

Levy, Hyman

Truth is a dangerous word to incorporate within the vocabulary of science. It drags with it, in its train, ideas of permanence and immutability that are foreign to the spirit of a study that is essentially an historically changing movement, and that relies so much on practical examination within restricted circumstances . . . Truth is an absolute notion that science, which is not concerned with any such permanency, had better leave alone.

The Universe of Science (pp. 206–7)

Newton, Sir Isaac

I do not know what I may appear to the world, but to myself I seem to have been only like a boy playing on the sea-shore, and diverting myself in now and then finding a smoother pebble or a prettier shell than ordinary, whilst the great ocean of truth lay all undiscovered before me.

In David Brewster
Memoirs of the Life, Writings and Discoveries of Sir Isaac Newton
Chapter 27 (p. 331)

Planck, Max

A new scientific truth does not triumph by convincing its opponents and making them see the light, but rather because its opponents eventually die, and a new generation grows up that is familiar with it.

Scientific Autobiography (pp. 33–4)

Reichenbach, Hans
He who searches for truth must not appease his urge by giving himself
up to the narcotic of belief.

In Ruth Renya
The Philosophy of Matter in the Atomic Era (p. 16)

Renan, Ernest
The simplest schoolboy is now familiar with truths for which Archimedes
would have sacrificed his life.

In L.I. Ponomarev
The Quantum Dice (p. 34)

Truth, sir, is a great coquette; she will not be sought with too much
passion, but often is most amenable to indifference. She escapes when
apparently caught, but gives herself up if patiently waited for; revealing
herself after farewells have been said, but inexorable when loved with
too much fervor.

In René Vallery-Radot
The Life of Pasteur
Volume II (p. 164)

Russell, Bertrand
When a man tells you that he knows the exact truth about anything, you
are safe in inferring that he is an inexact man.

The Scientific Outlook
Characteristics of Scientific Method (p. 65)

Spencer-Brown, George
To arrive at the simplest truth, as Newton knew and practiced, *requires*
years of *contemplation*. Not activity. Not reasoning. Not calculating. Not
busy behavior of any kind. Not reading. Not talking. Not making an
effort. Not thinking. Simply *bearing in mind* what it is one needs to know.
And yet those with the courage to tread this path to real discovery are not
only offered practically no guidance on how to do so, they are actively
discouraged and have to set about it in secret, pretending meanwhile to
be diligently engaged in the frantic diversions and to conform with the
deadening personal opinions which are continually being thrust upon
them.

Laws of Form
Appendix I (p. 110)

Wilde, Oscar
It is a terrible thing for a man to find out suddenly that all his life he
has been speaking nothing but the truth.

In John D. Barrow
The World within the World (p. 260)

JACK— . . . That, my dear Algy, is the whole truth, pure and simple.

ALGERNON—The truth is rarely pure and never simple.

The Importance of Being Earnest
Act I

Wilkins, John
That the strangeness of this opinion is no sufficient reason why it should be rejected, because other certain truths have been formerly esteemed ridiculous, and great absurdities entertayned by common consent.

The Discovery of a World in the Moone (p. 1)

The challenge for the modern architect is...to make out of the ordinary something out-of-the-ordinary.
Patrick Nuttgens – (See p. 7)

TUNNEL

Drinker, Henry
A barbarous people may, perhaps, develop a high degree of perfection in the mere art of open-air building, where stone can be piled on stone, and rafter fitted to rafter, in the light of day; but it takes the energy, knowledge, experience, and skill of an educated and trained class of men to cope with the unknown dangers of the dark depths that are to be invaded by the tunnel-man.

Tunneling, Explosive Compounds, and Rock Drilling (p. 32)

WEIGHT

da Vinci, Leonardo
Weight, pressure, and accidental movement together with resistance are
the four accidental powers in which all the visible works of mortals have
their existence and their end.

Leonardo da Vinci's Notebooks (p. 55)

BIBLIOGRAPHY

Abbott, Edwin A. *Flatland*. Barnes & Noble, Inc., New York. 1963.

Adams, Douglas. *Mostly Harmless*. Harmony Books, New York. 1992.

Adams, Franklin. *Tobogganing on Parnassus*. Doubleday, Page & Company, Garden City. 1918.

Adams, Henry. *The Education of Henry Adams*. 1946.

Adler, Alfred. *Atlantic Monthly*. Volume 279, Number 2. February 1997.

Akenside, Mark. *The Poetical Works of Mark Akenside and John Dyer*. George Routledge, New York. 1855.

Alcott, Louisa May. *Little Women*. The World Publishing Co., Cleveland. 1946.

Alger, John R.M. and Hays, Carl V. *Creative Synthesis in Design*. Prentice-Hall, Inc., Englewood Cliffs. 1964.

Alger, Philip L., Christensen, N.A. and Olmsted, Sterling P. *Ethical Problems in Engineering*. John Wiley & Sons, Inc., New York. 1965.

Allen, Roy George Douglas. *Statistics for Economists*. Hutchinson & Co., Ltd., London. 1949.

Amiel, Henri Frédéric. *Journal Intime*. A.L. Burt Company, Publisher. No date.

Anderson, Dale A., Tennehill, John C. and Pletcher, Richard H. *Computational Fluid Mechanics and Heat Transfer*. Hemisphere Publishing Corporation, Washington. 1984.

Andric, Ivo. *The Bridge on the Drina*. Translated by Lovett F. Edwards. George Allen & Unwin, London. 1959.

Anscombe, F.J. 'Rejection of Outliers' in *Technometrics*. Volume 2, Number 2. May 1960.

Aquinas, Thomas. *Summa Theologiae*. McGraw-Hill Book Co., New York. 1975.

Arber, Agnes. *The Mind and the Eye*. At the University Press, Cambridge. 1954.

Aristotle. *Nicomachean Ethics* in *Great Books of the Western World*. Translated by W.D. Ross. Volume 9. Encyclopædia Britannica, Inc., Chicago. 1952.

Aristotle. *On Generation and Corruption* in *Great Books of the Western World*. Translated by H.H. Joachim. Volume 8. Encyclopædia Britannica, Inc., Chicago. 1952.

Aristotle. *On Poetics* in *Great Books of the Western World*. Translated by Benjamin Jowett. Volume 9. Encyclopædia Britannica, Inc., Chicago. 1952.

Arm, David L. *Journeys in Science*. The University of New Mexico Press, Albuquerque. 1967.

Arthur, T.S. *Ten Nights in a Bar-Room and What I Saw There*. Edited by Donald A. Koch. The Belknap Press of Harvard University Press, Cambridge. 1964.

Arwaker, Edmund. *Pia Desidera*. Printed for Henry Bonwicke, London. 1686.

Ashby, Sir Eric. *Technology and the Academics*. Macmillan & Co., Ltd., London. 1959.

Asimov, Isaac. *Isaac Asimov's Book of Science and Nature Quotations*. Weidenfeld & Nicolson, New York. 1988.

Asimov, Isaac. *Of Time and Space and Other Things*. Avon Books, New York. 1965.

Asimow, Morris. *Introduction to Design*. Prentice-Hall, Inc., Englewood Cliffs. 1962.

Atherton, Gertrude. *Senator North*. John Wilson & Co., Cambridge. 1900.

Attwell, Henry. *Thoughts from Ruskin*. Longman, Green, and Co., New York. 1901.

Aurelius, Marcus. *The Meditations of the Emperor Antoninus Marcus Aurelius*. Translated by George Long. Thomas Y. Crowell & Co., New York. No date.

Ayres, C.E. *Science: The False Messiah*. The Bobbs-Merrill Company, Indianapolis. 1927.

Bacon, Francis. *Novum Organum* in *Great Books of the Western World*. Volume 30. Encyclopædia Britannica, Inc., Chicago. 1952.

Bacon, Francis. *The Advancement of Learning* in *Great Books of the Western World*. Volume 30. Encyclopædia Britannica, Inc., Chicago. 1952.

Bacon, Roger. *The Opus Majus of Roger Bacon*. Volume 1. A translation by Robert Belle Burke. Russell & Russell, Inc., New York. 1962.

Baez, Joan. *Daybreak*. Avon Books, New York. 1968.

Bagehot, Walter. *Physics and Politics*. D. Appleton and Company, New York. 1884.

Bailey, Philip James. *Festus*. George Routledge and Sons, Limited, Manchester. 1893.

Bailey, Robert I. *Disciplined Creativity for Engineers*. Ann Arbor Science Publishers Inc., Ann Arbor. 1978.

Baillie, Joanna. *Miscellaneous Plays*. Longman, Hurst, Rees, and Orme, London. 1805.

Balchin, Nigel. *The Small Back Room*. Collins, London. 1943.

Ballard, J.G. *Crash*. Flamingo, London. 1993.

Barr, H.F. 'Typical Problems in Engineering' in *General Motors Engineering Journal*. Set 1, Number 1.

Barrow, John D. *The Artful Universe*. Clarendon Press, Oxford. 1995.

Barrow, John D. *The World within the World*. Clarendon Press, Oxford. 1988.

Barry, Frederick. *The Scientific Habit of Thought*. Columbia University Press, New York. 1927.

Bartol, C.A. *Radical Problems*. Robert Brothers, Boston. 1872.

Bartusiak, Marcia. *Thursday's Universe*. Times Books, New York. 1986.

Beakley, George C. *Careers in Engineering and Technology*. Macmillan Publishing Company, New York. 1984.

Bean, William B. *Aphorisms from Latham*. Prairie Press, Iowa City. 1962.

Beard, George M. 'Experiments with Living Human Beings' in *Popular Science Monthly*. Volume 14. 1879.

Beer, Stafford. *Brain of the Firm*. John Wiley & Sons, Chichester. 1981.

Bell, Eric T. *Men of Mathematics*. Simon and Schuster, New York. 1937.

Bell, Eric T. *The Search for Truth*. George Allen & Unwin Ltd., London. 1946.

Bellman, Richard. *Eye of the Hurricane: An Autobiography*. World Scientific, Singapore. 1984.

Bergson, Henri. *Creative Evolution*. Henry Holt and Company, New York. 1911.

Berkeley, Edmund C. 'Right Answers—A Short Guide for Obtaining Them' in *Computers and Automation*. September 1969.

Bernard, Claude. *An Introduction to the Study of Experimental Medicine*. H. Schuman, New York. 1927.

Beveridge, W.I.B. *The Art of Scientific Investigation*. Norton, New York. 1957.

Beyer, Robert W. 'Humor Rumor: It Helps' in *The Physics Teacher*. December 1977.

Bierce, Ambrose. *The Enlarged Devil's Dictionary*. Doubleday & Company, Inc., Garden City. 1967.

Billings, Josh. *Old Probability: Perhaps Rain—Perhaps Not*. Literature House, Upper Saddle River. 1970.

Billington, David P. *The Tower and the Bridge*. Basic Books, Inc., New York. 1983.

Billington, David P. 'The Defense of Engineers' in *Wilson Quarterly*. Volume 10, Number 1. 1986.

The Tower and the Bridge. Basic Books, Inc., New York. 1983.

Birns, Harold. 'City Acts to Unify Inspection Rules' in *New York Times*. October 2, 1963.

Blackwood, Oswald. *Introductory College Physics*. John Wiley & Sons, Inc., New York. 1939.

Blake, William. *The Prophetic Writings of William Blake*. At the Clarendon Press, Oxford. 1926.

Bloch, Arthur. *Murphy's Law*. Price/Stern/Sloan Publishers, Inc., Los Angeles. 1979.

Bohr, Niels. *Atomic Theory and the Description of Nature*. At the University Press, Cambridge. 1961.

Bolton, Arthur T. *Life and Work A Century Ago: An Outline of the Career of Sir John Soane*. Soane Museum Publications, Number 11. 1923?

Bondi, Hermann. *Relativity and Common Sense*. Doubleday & Company, Inc., Garden City. 1964.

Boorstin, Daniel J. *Hidden History*. Vintage Books, New York. 1989.

Borel, Émile. *Probability and Certainty*. Walker, New York. 1963.

Borges, Jorge Luis. *Conversations with Jorge Luis Borges*. Holt, Rinehart and Winston, New York. 1968.

Born, Max. *The Restless Universe*. Dover Publications, New York. 1951.

Boswell, James. *The Life of Samuel Johnson*. International Collectors Library, Garden City. 1945.

Bottomley, Gordon. *Poems of Thirty Years*. Constable & Company Limited, London. 1925.

Bowen, Elizabeth. *The Death of the Heart*. Alfred A. Knopf, New York. 1939.

Box, G.E.P. 'Use and Abuse of Regression' in *Technometrics*. Volume 8, Number 4. November 1966.

Boyd, T.A. *Professional Amateur*. E.P. Dutton & Co., Inc., New York. 1957.

Bradley, A.C. *Oxford Lectures on Poetry*. Macmillan and Company, Limited, London. 1909.

Bradley, Duane. *Engineers Did It!* J.B. Lippincott Company, Philadelphia. 1958.

Bragg, William L. 'The Physical Sciences' in *Science*. March 16, 1934.

Brecht, Bertolt. *The Messingkauf Dialogues*. Translated by John Willett. Methuen & Co., Ltd., London. 1963.

Brewster, Sir David. *Memoirs of the Life, Writings, and Discoveries of Sir Isaac Newton*. Volume 2. Edinburgh, Edmonston. 1860.

Bridges, Robert Seymour. *The Growth of Love*. Thomas B. Mosher, Portland. 1913.

Bronowski, Jacob. *The Ascent of Man*. Little, Brown and Company, Boston. 1973.

Bronowski, J. *The Common Sense of Science*. Harvard University Press, Cambridge. 1953.

Bronowski, Jacob. 'Technology and Culture in Evolution' in *The American Scholar*. Volume 41, Number 2. Spring 1972.

Broun, Heywood. *Seeing Things at Night*. Harcourt, Brace and Company, New York. 1921.

Brown, Allan. *Golden Gate*. Doubleday & Company, Inc., Garden City. 1965.

Brown, Gordon S. 'Engineer–Scientist' in *Electronics*. Volume 32, Number 47. November 20, 1959.

Brown, Gorden Spencer. *Laws of Form*. Allen & Unwin, London. 1969.

Brown, Kenneth A. *Inventors at Work*. Tempus Books of Microsoft Press, Redmond. 1988.

Buchanan, R. Angus. *Engineers and Engineering*. Bath University Press, Calverton Down. 1996.

Buchanan, R.W. *The Complete Poetical Works of Robert Buchanan*. Chatto & Windus, London. 1901.

Buchanan, Scott. *Poetry and Mathematics*. J.B. Lippincott Company, Philadelphia. 1962.

Buckles, Robert A. *Ideas, Inventions, and Patents*. John Wiley & Sons, Inc., New York. 1957.

Bugliarello, G. and Doner, D.B. *The History and Philosophy of Technology*. University of Illinois Press, Urbana. 1979.

Bullock, James. 'Literacy in the Language of Mathematics' in *American Mathematical Monthly*. Volume 101, Number 8. October 1994.

Burton, Robert. *Anatomy of Melancholy*. Clarendon Press, Oxford. 1989.

Bush, Vannevar. *Endless Horizons*. Public Affairs Press, Washington, D.C. 1946.

Butler, Samuel. *Samuel Butler's Notebooks*. Edited by Geoffrey Keyner and Brian Hill. Jonathan Cape, London. 1951.

Byron, Lord George Gordon. *The Poetical Works of Lord Byron*. Oxford University Press, New York. 1946.

Cage, John. *Silence 1961*. Wesleyan University Press, Middletown. 1961.

Campbell, Lewis and Garnett, William. *The Life of James Clerk Maxwell*. Macmillan and Company, London. 1882.

Camus, Albert. *The Fall*. Vintage Books, New York. 1956.

Cannon, Walter Bradford. *The Way of an Investigator*. W.W. Norton & Company, Inc., New York. 1945.

Cardenal, Ernesto. *Cosmic Canticle*. Curbstone Press, Willimantic. 1993.

Cardozo, Benjamin N. *The Paradoxes of Legal Science*. Columbia University Press, New York. 1927.

Cardwell, D.S.L. 'Technology' in *Dictionary of the History of Ideas*. Volume IV. Edited by Philip P. Wiener. Charles Scribner's Sons, New York. 1973.

Carlyle, Thomas. *English and Other Critical Essays*. J.M. Dent & Sons Ltd., London. 1950.

Carlyle, Thomas. *Sartor Resartus*. Frederick A. Stokes Company, Publisher, New York. 1893.

Carroll, Lewis. *The Complete Works of Lewis Carroll*. Modern Library, New York. 1936.

Carr-Saunders, A.M. and Wilson, P.A. *The Professions*. At the Clarendon Press, Oxford. 1933.

Casimir, Hendrik. *Haphazard Reality*. Harper & Row, Publishers, New York. 1983.

Cervantes, Miguel de. *Don Quixote* in *Great Books of the Western World*. Volume 29. Encyclopædia Britannica, Inc., Chicago. 1952.

Chargaff, Erwin. 'Bitter Fruits from the Tree of Knowledge' in *Perspectives in Biology and Medicine*. Volume 16. Summer 1973.

Chargaff, Erwin. 'In Praise of Smallness' in *Perspectives in Biology and Medicine*. Spring 1980.

Chaucer, Geoffrey. *Troilus and Cressida* in *Great Books of the Western World*. Translated by George Philip Krappt. Volume 22. Encyclopædia Britannica, Inc., Chicago. 1952.

Chesterton, G.K. *The G.K. Chesterton Calendar*. Cecil Palmer & Hayward, London. 1916.

Chesterton, G.K. *The Scandal of Father Brown*. Dodd, New York. 1935.

Clarke, Arthur C. *2001*. William Heinemann Limited, London. 1985.

Clarke, Arthur C. *The Lost Worlds of 2001*. Gregg Press, Boston. 1972.

Clarke, J.M. 'Overhead Costs in Modern Industry' in *Journal of Political Economy*. October 1927.

Clemence, G.M. 'Time and its Measurement' in *The American Scientist*. Volume 40, Number 2. April 1952.

Cohen, I. Bernard. *The Birth of A New Physics*. Norton, New York. 1985.

Cohen, Morris. *A Preface to Logic*. H. Holt and Company, New York. 1944.

Cohen, Morris R. and Nagel, Ernest. *An Introduction to Logic and Scientific Method*. Harcourt, Brace and Company, New York. 1934.

Colclaser, R.G. 'A Design School for the Young Engineer in Industry' in *Engineering Education*. March 1968.

Collingwood, Robin George. *Speculum Mentis*. The Clarendon Press, Oxford. 1924.

Colton, Charles Caleb. *Lacon*. William Gowans, New York. 1849.

Colvin, H.M. *A Biographical Dictionary of English Architects, 1600–1840*. John Murray (Publishers) Ltd., London. 1978.

Commoner, Barry. *Science and Survival*. The Viking Press, New York. 1966.

Compton, Karl Taylor. *A Scientist Speaks*. Massachusetts Institute of Technology, Cambridge. 1955.

Compton, Karl Taylor. 'Engineering and Social Progress' in *The Journal of Engineering Education*. Volume 30, Number 1. September 1939.

Cort, David. *Social Astonishments*. The Macmillan Company, New York. 1963.

Cowper, William. *The Complete Poetical Works of William Cowper*. Oxford University Press, London. 1913.

Cresy, Edward. *A Practical Treatise on Bridge Building and on the Equilibrium of Vaults, and Arches with the Professional life and Selections from the Works of Rennie*. 1839.

Crew, Henry. *General Physics*. The Macmillan Company, New York. 1927.

Crichton, Michael. *Jurassic Park*. Alfred A. Knopf, New York. 1990.

Crick, Francis. *What Mad Pursuit*. Basic Books, Inc., Publishers, New York. 1988.

Cross, Hardy. *Engineers and Ivory Towers*. McGraw-Hill Book Company, Inc., New York. 1952.

Czarnomski, F.B. *The Wisdom of Winston Churchill*. George Allen and Unwin Ltd., London. 1956.

da Vinci, Leonardo. *Leonardo da Vinci's Notebooks*. Duckworth & Co., London. 1906.

da Vinci, Leonardo. *The Literary Works of Leonardo da Vinci*. Volume I. Compiled and edited by Jean Paul Richter. Oxford University Press, London. 1937.

da Vinci, Leonardo. *The Literary Works of Leonardo da Vinci*. Volume II. Compiled and edited by Jean Paul Richter. Oxford University Press, London. 1937.

da Vinci, Leonardo. *The Notebooks of Leonardo da Vinci*. Volume I. Translated by Edward MacCurdy. Reynal & Hitchcock, New York. 1938.

Darrow, Karl K. *The Renaissance of Physics*. The Macmillan Company, New York. 1936.

Darwin, Francis. *The Life and Letters of Charles Darwin*. Volume II. D. Appleton and Company, New York. 1896.

Davies, Paul. *About Time*. Simon & Schuster, New York. 1995.

Davies, Paul. *Other Worlds*. Simon and Schuster, New York. 1980.

de Beauvoir, Simone. *The Mandarins*. The World Publishing Co., Cleveland. 1956.

de Bergerac, Cyrano. *The Comical History of the States and Empires of the Worlds of the Moon and Sun*. Translated by A. Lovell. Printed for *Henry Rhodes*, next door to the *Swan Tavern*, near *Bride Lane*, in *Fleet Street*, London. 1687.

de Camp, L. Sprague. *The Ancient Engineers*. Doubleday & Company, Inc., Garden City. 1963.

de Costa, J. Chalmers. *The Trials and Triumphs of the Surgeon*. Dorrance, Philadelphia. 1944.

de Saint-Exupéry, Antoine. *Wind, Sand and Stars*. Harcourt, Brace and Company, Inc., New York. 1940.

Dean, Thomas C. 'Challenges in Higher Education' in *Phi Kappa Phi Journal*. Volume L, Number 3. Summer 1970.

Defoe, Daniel. *Selected Poetry and Prose of Daniel Defoe*. Holt, Rinehart and Winston, New York. 1968.

Deming, William Edwards. *Some Theory of Sampling*. John Wiley & Sons, Inc., New York. 1960.

Deming, William Edwards. *Statistical Adjustment of Data*. Chapman & Hall, Ltd., London. 1943.

Deming, William Edwards. 'On a Classification of the Problems of Statistical Inference' in *Journal of the American Statistical Association*. Volume 37, Number 218. June 1942.

Deming, William Edwards. 'On the Classification of Statistics' in *The American Statistician*. Volume 2, Number 2. April 1948.

DeSimone, Daniel V. *Education for Innovation*. Pergamon Press, Ltd., Elmsford. 1968.

Dewey, John. *Logic: The Theory of Inquiry*. Henry Holt and Company, New York. 1938.

Dewey, John. *Reconstruction in Philosophy*. H. Holt & Company, New York. 1920.

Dewey, John. *The Quest for Certainty*. Minton, Balch & Company, New York. 1929.

Diamond, Solomon. *Information and Error*. Basic Books, Inc., New York. 1959.

Dibdin, Charles Isaac Mungo. *Young Arthur*. Printed by Strahan and Spotiswodde, London. 1819.

Dickens, Charles. *The Pickwick Papers*. Chapman & Hall, London. 1836.

Dickinson, Emily. *Further Poems*. Little, Brown, and Company, Boston. 1929.

Dickinson, J.C. *The Great Charter, A Translation*. Published for the Historical Association by G. Philip, London. 1955.

Dieudonné, Jean. *Mathematics—The Music of Reason*. Springer-Verlag, Berlin. 1987.

Dimnet, Ernest. *What We Live By*. Simon and Schuster, New York. 1932.

Dirac, Paul Adrien Maurice. 'The Evolution of the Physicist's Picture of Nature' in *Scientific American*. Volume 208, Number 5. May 1963.

Disraeli, Benjamin. *Coningsby on the New Generation*. Penguin Books Ltd., Harmondsworth. 1983.

Disraeli, Isaac. *Literary Characters of Men of Genius*. F. Warne, London. 1881?

Dobie, J. Frank. *The Voice of the Coyote*. Little, Brown and Company, Boston. 1949.

Dobzhansky, Theodosius. *The Biology of Ultimate Concern*. The New American Library, Inc., New York. 1967.

Donghia, Angelo. 'Behind Angelo Donghia's Gray Flannel Success' in *New York Times*. January 20, 1983.

Doyle, Arthur Conan. *The Complete Sherlock Holmes*. Garden City Publishing Co., New York. 1930.

Drachmann, A.G. *The Mechanical Technology of Greek and Roman Antiquity*. Munksgaard, Copenhagen. 1963.

Drexler, K. Eric. *Engines of Creation*. Anchor Press, Garden City. 1986.

Drinker, Henry. *Tunneling, Explosive Compounds, and Rock Drilling*. John Wiley & Sons, New York. 1888.

Drummond, Sir William. *Academical Questions*. Scholars' Facsmiles & Reprints, Delmar. 1984.

Dryden, John. *The Poetical Works of Dryden*. Macmillan, New York. 1904.

Du Noüy, Pierre L. *The Road to Reason*. Longmans & Green, Toronto. 1949.

Dubos, René. *A God Within*. Charles Scribner's Sons, New York. 1972.

Dudley, Underwood. 'Formulas for Primes' in *Mathematical Magazine*. Volume 56, 1983.

Duhem, Pierre. *The Aim and Structure of Physical Theory*. Atheneum, New York. 1977.

Dunsany, Lord Edward John Moreton Drax Plunkett. *My Ireland*. Jarrold Publishers, London. 1937.

Dyer, Frank Lewis and Martin, Thomas Commerford. *Edison: His Life and Inventions*. Volume II. Harper & Brothers Publishers, New York. 1929.

Dylan, Bob. 'The Two Lives of Bob Dylan' in *Newsweek*. 8 December. 1985.

Dyson, Freeman. *Disturbing the Universe*. Harper & Row, Publishers, New York. 1979.

Dyson, Freeman. 'Energy in the Universe' in *Scientific American*. Volume 224, Number 3. September 1971.

Easton, Elmer C. 'An Engineering Approach to Creative Thinking' in *Ceramic Age*. September 1955.

Eckermann, Johann Peter. *Conversations with Goethe*. Dent, London. 1971.

Eddington, Sir Arthur Stanley. *The Philosophy of Physical Science*. Cambridge University Press, London. 1949.

Edwards, Llewellyn Nathaniel. *A Record of History and Evolution of Early American Bridges*. University Press, Orono. 1959.

Edwards, Tyrone. *The New Dictionary of Thoughts*. Standard Book Company, New York. 1955.

Egerton, Sarah. *Poems on Several Occasions*. Scholar's Facsimiles & Reprints, Delmar. 1987.

Ehrenberg, A.S.C. *Data Reduction*. John Wiley & Sons, New York. 1975.

Einstein, Albert. *Out of My Later Years*. Thames and Hudson, London. 1950.

Einstein, Albert. *The Evolution of Physics*. Simon and Schuster, New York. 1938.

Einstein, Albert. *The World As I See It*. Philosophical Library, New York. 1949.

Eliot, George. *Daniel Deronda*. W. Blackwood, Edinburgh. 1876.

Emerson, Ralph Waldo. *English Traits, Representative Men, Other Essays*. J.M. Dent & Sons Ltd., London. 1951.

Emerson, Ralph Waldo. *Essays*. Oxford University Press, London. 1901.

Emerson, Ralph Waldo. *Letters and Social Aims*. J.R. Osgood and Company, Boston. 1876.

Emerson, Ralph Waldo. *Society and Solitude*. Fields, Osgood and Co., Boston. 1870.

Emerson, Ralph Waldo. *The Works of Ralph Waldo Emerson*. Volumes I, III and VI. Boston. 1909.

Emme, Eugene M. 'Introduction to the History of Rocket Technology' in *Technology and Culture*. Fall 1963.

Emmerson, G.S. *Engineering Education: A Social History*. David & Charles, New York. 1973.

Emmet, William Le Roy. *The Autobiography of an Engineer*. The American Society of Mechanical Engineers, New York. 1940.

Epstein, P.S. *Textbook of Thermodynamics*. John Wiley & Sons, Inc., New York. 1937.

Esar, Evan. *20,000 Quips and Quotes*. Barnes & Noble Books, New York. 1968.

Evans, Bergen. *The Natural History of Nonsense*. A.A. Knopf, New York. 1946.

Faber, Harold. *The Book of Laws*. Times Books, New York. 1979.

Fabing, Harold and Marr, Ray. *Fischerisms*. C.C. Thomas, Springfield. 1937.

Faraday, Michael. *On the Various Forces of Nature*. Thomas Y. Crowell & Co., New York. 1957.

Farber, Eric A. 'The Teaching and Learning of Engineering' in *Journal of Engineering Education*. Volume 45, Number 10. June 1955.

Farquhar, George. *The Beaux' Stratagem*. The Scolar Press, Limited, London. 1972.

Ferguson, Eugene S. *Engineering and the Mind's Eye*. The MIT Press, Cambridge. 1992.

Fergusson, James. *History of Indian and Eastern Architecture*. Munshiram Manoharlal, Nai Sark. 1967.

Feynman, Richard. *The Character of Physical Law*. British Broadcasting Corporation, London. 1965.

Feynman, Richard P. *What Do You Care What Other People Think?* W.W. Norton & Company, New York. 1988.

Feynman, Richard P., Leighton, Robert B. and Sands, Matthew. *The Feynman Lectures on Physics*. Volume 1. Addison-Wesley Publishing Company, Reading. 1964.

Fielder, Edgar R. 'The Three Rs of Economic Forecasting—Irrational, Irrelevant and Irreverent' in *Across the Board*. June 1977.

Finch, James K. *Engineering and Western Civilization*. McGraw-Hill Book Company, Inc., New York. 1951.

Finch, James Kip. *The Story of Engineering*. Doubleday & Company, Inc., Garden City. 1960.

Fish, John Charles Lounsbury. *Engineering Economics: First Principles*. McGraw-Hill Book Company, Inc., New York. 1915.

Fish, John Charles Lounsbury. *The Engineering Method*. Stanford University Press, Stanford. No date.

Fisher, Irving. *The Theory of Interest*. The Macmillan Company, New York. 1930.

Fitzgerald, F. Scott. *The Crack-Up*. James Laughlin, New York. 1945.

Flinn, Alfred D. 'Leadership in Economic Progress' in *Civil Engineer*. Volume 2, Number 4. April 1932.

Florman, Samuel C. *Blaming Technology*. St. Martin's Press, New York. 1981.

Florman, Samuel C. 'Comment: Engineers and the End of Innocence' in *Technology and Culture*. Volume 10, Number 1, January 1969.

Florman, Samuel C. *Engineering and the Liberal Arts*. McGraw-Hill Book Company, New York. 1968.

Florman, Samuel C. *The Civilized Engineer*. St. Martin's Press, New York. 1987.

Florman, Samuel C. *The Existential Pleasures of Engineering*. St. Martin's Press, New York. 1976

Fox, Russell, Gorbuny, Max and Hooke, Robert. *The Science of Science*. Walker and Company, New York. 1963.

Franklin, Benjamin. *The Works of Benjamin Franklin*. Volume I. Longman, Hurst, Rees, & Orme, London. 1806.

Frazier, A.W. 'The Practical Side of Creativity' in *Hydrocarbon Processing*. Volume 45, Number 1. January 1966.

Fredrickson, A.G. 'The Dilemma of Innovating Societies' in *Chemical Engineering Education*. Volume 4. Summer 1969.

French, A.P. *Einstein: A Centenary Volume*. Harvard University Press, Cambridge. 1979.

Freud, Sigmund. *On Narcissism* in *Great Books of the Western World*. Translated by Cecil M. Baines. Volume 54. Encyclopædia Britannica, Inc., Chicago. 1952.

Freund, C.J. 'Creativity is a Task, Not a Trait' in *Machine Design*. May 25, 1967.

Freund, C.J. 'Engineering Education and Freedom from Fear' in *The Journal of Engineering Education*. Volume 40, Number 1. September 1949.

Freyssinet, E. *The Birth of Prestressing*. Translated from Travaux (Paris). Library translation No. 59. Cement and Concrete Association, London. 1956.

Friedel, Robert. 'Engineering in the 20th Century' in *Technology and Culture*. Special Issue. October 1968.

Friedman, Milton and Friedman, Rose. *Free to Choose*. Harcourt Brace Jovanovich, New York. 1980.

Froude, James Anthony. *Short Studies on Great Subjects*. E.P. Dutton, New York. 1930–31.

Fuller, Thomas. *Gnomologia*. Printed by S. Powell, Dublin. 1733.

Gabor, Dennis. *Innovations: Scientific, Technological, and Social*. Oxford University Press, London. 1970.

Gabor, Dennis. *Inventing the Future*. Secker & Warburg, London. 1963.

Galbraith, John Kenneth. *The Affluent Society*. The New American Library, New York. 1958.

Galbraith, John Kenneth. *The New Industrial State*. Houghton Mifflin Company, Boston. 1985.

Galilei, Galileo. 'Dialogues Concerning the Two New Sciences' in *Great Books of the Western World*. Translated by Henry Crew and Alfonso de Salvio. Encyclopædia Britannica, Inc., Chicago. 1952.

Gall, John. *Systemantics*. Pocket Books, New York. 1977.

Galsworthy, John. *Eldest Son*. Charles Scribner's Sons, New York. 1912.

Gardner, Earl Stanley. *The Case of the Perjured Parrot*. The Blakiston Company, Philadelphia. 1939.

Gardner, Martin. *Fads and Fallacies*. Dover Publications, Inc., New York. 1957.

Gardner, M. *Great Essays in Science*. Prometheus Books, Buffalo. 1984.

Garfield, E. 'Creativity and Science' in *Current Comments*. Number 43. October 23, 1989.

Garrison, F.H. *Bulletin of the New York Academy of Medicine*. Volume 4. 1928.

Good, Irving John. *The Scientist Speculates*. Basic Books, Inc., Publishers, New York. 1962.

Gordon, James E. *The New Science of Strong Materials*. Walker and Company, New York. 1968.

Graham, L.A. *Ingenious Mathematical Problems and Methods*. Dover Publications, Inc., New York. 1959.

Green, Celia. *The Decline and Fall of Science*. Hamilton, London. 1976.

Greenstein, Jesse L. 'Great American Scientists: The Astronomers' in *Fortune*. Volume 61, Number 5. May 1960.

Greer, Scott. *The Logic of Social Inquiry*. Aldine Publication Co., Chicago. 1969.

Gregg, Alan. *The Furtherance of Medical Research*. Yale University Press, New Haven. 1941.

Gregory, Olinthus. 'A Treatise of Mechanics' in *American Journal of Science*. Volume 7. 1824.

Grinter, L.E. 'Report on Evaluation of Engineering Education (1952–1955)' in *Journal of Engineering Education*. Volume 46, Number 1. September 1955.

Gruenberg, Benjamin C. *Science and the Public Mind*. McGraw-Hill Book Company, Inc., New York. 1935.

Gudder, Stanley. *A Mathematical Journey*. McGraw-Hill Book Company, New York. 1976.

Guest, Edgar A. *The Path Home*. The Reilly & Lee Co., Chicago. 1919.

Gunther, John. *Inside Russia Today*. Harper & Brothers, New York. 1958.

Hacking, Ian. *The Emergence of Probability*. Cambridge University Press, Cambridge. 1975.

Hamilton, L.L. 'Engineer Report' in *American Engineer*. Volume 29, Number 6. June 1959.

Hammond, A.L. and Metz, W.D. 'Solar Energy Research: Making Solar After the Nuclear Model?' in *Science*. Volume 197, 1977.

Hammond, H.P. 'Engineering Education After the War' in *Journal of Engineering Education*. Volume 43, Number 9. May 1944.

Hammond, H.P. 'Report of Committee on Aims and Scope of Engineering Curricula' in *Journal of Engineering Education*. Volume 30, Number 7. March 1940.

Hammurabi. *The Codes of Hammurabi and Moses*. By W.W. Davies. The Methodist Book Concern, New York. 1905.

Harrington, Eldred. *An Engineer Writes about People and Places and Projects*. Calvin Horn, Publishers, Inc., Albuquerque. No date.

Harrington, Michael. *The Other America*. Penguin Books, New York. 1971.

Harris, A.J. 'Architectural Misconceptions of Engineering' in *Journal of the Royal Institute of British Architects*. Volume 68, 3rd Series.

Harris, Ralph. *Economic Forecasting—Models or Markets?* The Institute of Economic Affairs, London. 1977.

Hawking, Stephen. *A Brief History of Time*. Bantam Books, New York. 1988.

Heaviside, Oliver. *Electromagnetic Theory*. Dover Publications, Inc., New York. 1950.

Heidegger, Martin. *Basic Writings*. Harper and Row, Publishers, San Francisco. 1977.

Heinlein, Robert A. *Expanded Universe*. Grosset & Dunlap Publishers, New York. 1980.

Heinlein, Robert A. *The Cat Who Walks Through Walls*. Putnam, New York. 1985.

Heisenberg, Werner. *Physics and Beyond: Encounters and Conversations*. Harper and Row, New York. 1971.

Heisenberg, Werner. *Physics and Philosophy*. Harper Torchbooks, New York. 1962.

Hembree, Lawrence. *Quote*. Volume 54, Number 6. April 6, 1967.

Herbert, George. *The Country Parson, The Temple*. Paulist Press, New York. 1981.

Hersey, John. *A Single Pebble*. Alfred A. Knopf, New York. 1956.

Herstein, I.N. *Topics in Algebra*. John Wiley & Sons, New York. 1975.

Hertz, H. *Electric Waves; Being Researches on the Propagation of Electric Action with Finite Velocity through Space*. Macmillan & Co., London. 1893.

Heywood, Robert B. *The Works of the Mind*. The University of Chicago Press, Chicago. 1947.

Hilbert, David. 'Hilbert: Mathematical Problems' in *Bulletin of the American Mathematical Society*. Volume 8, 2nd Series. July 1902.

Hill, A.V. *The Ethical Dilemma of Science and Other Writings*. The Rockefeller Institute Press, New York. 1960.

Hodnett, Edward. *The Art of Problem Solving*. Harper & Brothers, Publishers, New York. 1955.

Hoefler, Don C. 'But You Don't Understand the Problem' in *Electronic News*. July 17, 1967.

Hoffer, Eric. *The Passionate State of Mind*. Harper & Brothers, New York. 1955.

Hogben, Lancelot. *Science for the Citizen*. Alfred A. Knopf, New York. 1938.

Holmes, Oliver Wendell. *The Autocrat of the Breakfast-Table*. Houghton Mifflin, Boston. 1894.

Holmes, Oliver Wendell. *The Complete Poetical Works of Oliver Wendell Holmes*. Houghton, Mifflin and Co., Boston. 1881.

Holmes, Oliver Wendell. *The Poet at the Breakfast-Table*. James R. Osgood and Company, Boston. 1872.

Holmes, Oliver Wendell. *The Professor at the Breakfast-Table*. Houghton, Mifflin and Company, Boston. 1891.

Hood, Thomas. *The Comic Poems of Thomas Hood*. E. Moxon & Co., London. 1868.

Hooke, Robert. *Micrographia*. Weinheim, Cramer, New York. 1961.

Hoover, Herbert. *The Memoirs of Herbert Hoover*. Volume I. The Macmillan Company, New York. 1952.

Howland, W.E. and Wiley, R.B. 'Backsight at a Turning Point' in *Civil Engineering*. Volume XI, Number 4. April 1941.

Hubbard, Elbert. *Notebook*. Wm. H. Wise & Co., New York. 1927.

Hugo, Victor. *Intellectual Autobiography*. Translated by L. O'Rourke. Funk & Wagnalls Co., London. 1907.

Hugo, Victor. *Les Misérables*. Translated by Isabel F. Hapgood. T.Y. Crowell & Co., New York. 1887.

Hume, David. *An Enquiry Concerning Human Understanding*. The Open Court Publishing Co., Chicago. 1921.

Hutchins, Robert M. and Adler, Mortimer J. *The Great Ideas Today, 1974*. Encyclopædia Britannica, Inc., Chicago. 1974.

Hutten, Ernest H. *The Language of Modern Physics*. George Allen and Unwin, Ltd., London. 1956.

Huxley, Aldous. *Brave New World*. Harper & Row, Publishers, New York. 1946.

Huxley, Aldous. *Literature and Science*. Harper & Row, New York. 1963.

Huxley, Aldous. *Music at Night and Other Essays*. The Fountain Press, New York. 1931.

Huxley, Aldous. *Proper Studies*. Chatto & Windus, London. 1949.

Huxley, Aldous. *Science, Liberty and Peace*. Harper & Brothers Publishers, New York. 1946.

Huxley, Aldous. *Texts and Pretexts*. Greenwood Press Publishers, Westport. 1961.

Huxley, Aldous. *Tomorrow and Tomorrow and Tomorrow*. Harper & Row, New York. 1972.

Huxley, Julian. *Essays in Popular Science*. Alfred A. Knopf, New York. 1927.

Huxley, Thomas. *Collected Essays*. Volume II. Macmillan and Co., Limited, London. 1907.

Huxley, Thomas. *Huxley's Essays*. The Macmillan Company, New York. 1938.

Huxley, Thomas. *Lay Sermons, Addresses, and Reviews*. D. Appleton & Company, New York. 1871.

Huxley, Thomas. *Man's Place in Nature*. University of Michigan Press, Ann Arbor. 1959.

Huxley, Thomas. *Methods and Results*. D. Appleton and Co., New York. 1896.

Huxley, Thomas. *Science and Education*. A.L. Fowle, New York. 1893.

Huxley, Thomas. *The Life and Letters of Thomas Huxley*. D. Appleton, New York. 1901.

Huxley, Thomas. 'Scientific Education: Notes of an After Dinner Speech' in *Macmillian's Magazine*. Volume XX. July 1869.

James, William. *Varieties of Religious Experience*. Modern Library, New York. 1929.

James, William. 'The Sentiment of Rationality' in *Mind*. Series 1, Volume 4. 1879.

Jeans, Sir James. *The Mysterious Universe*. At the University Press, Cambridge. 1930.

Jefferson, Thomas. *Notes on the State of Virginia*. J.W. Randolph, Richmond. 1853.

Jerome K. Jerome. *The Idle Thoughts of an Idle Fellow*. H. Altemus, Philadelphia. 1890.

Jevons, W.S. *The Principles of Science*. Dover Publication, Inc., New York. 1958.

Johnson, George. *Fire in the Mind*. Alfred A. Knopf, New York. 1995.

Johnson, James Weldon. *Fifty Years & Other Poems*. The Cornhill Company. 1917.

Johnson, Philip. 'Ideas and Men' in *New York Times*. 27 December 1964.

Johnson, Samuel. *Rasselas*. A.L. Bart & Company, Publishers, New York. 1870.

Jones, Raymond F. *The Non-Statistical Man*. Belmont Productions, Inc., New York. 1964.

Jonson, Ben. *Volpone*. Yale University Press, New Haven. 1962.

Joseph, Margaret. 'The Future of Geometry' in *The Mathematics Teacher*. Volume XXIX, Number 1. January 1936.

Juster, Norton. *The Phantom Tollbooth*. Epstein & Carroll Associates, Inc., New York. 1962.

Kaempffert, Waldemar. *Invention and Society*. American Library Association, Chicago. 1930.

Kafka, Franz. *The Castle*. Alfred A. Knopf, New York. 1943.

Kaplan, Abraham. *The Conduct of Inquiry*. Chandler Publishing Co., San Francisco. 1964.

Kasner, Edward and Newman, James. *Mathematics and the Imagination*. Simon and Schuster, New York. 1967.

Kaysen, Carl. 'Limits to Growth' in *Foreign Affairs*. Volume 50, Number 4. July 1972.

Keate, George. *The Distressed Poet*. Printed for J. Dodsley, London. 1787.

Keats, John. *The Poetical Works of John Keats*. Oxford University Press, London. 1946.

Keeney, Ralph L. and Raiffa, Howard. *Decisions with Multiple Objectives: Preferences and Value Tradeoffs*. John Wiley & Sons, New York. 1976.

Kelly-Bootle, Stan. *The Devil's DP Dictionary*. McGraw-Hill Book Co., Inc., New York. 1981.

Kettering, C.F. *Professional Amateur*. Dutton, New York. 1957.

Kettering, C.F. *Short Stories of Science and Invention*. General Motors, Detroit. 1955.

Keyser, Cassius J. *Mathematical Philosophy: A Study of Fate and Freedom*. E.P. Dutton & Company, New York. 1922.

Keyser, Cassius J. *Mole Philosophy & Other Essays*. E.P. Dutton & Company, New York. 1927.

King, Blake. 'Object: Creativity' in *Mechanical Engineering*. November 1963.

King, W.J. 'The Unwritten Laws of Engineering' in *Mechanical Engineering*. June 1944.

Kingsley, Charles. *The Works of Charles Kingsley*. Volume II. Macmillan and Co., London. 1881.

Kipling, Rudyard. *Collected Verse of Rudyard Kipling*. Doubleday, Page & Company, Garden City. 1925.

Kipling, Rudyard. *From Sea to Sea*. Doubleday, Page & Co., Garden City. 1913.

Kline, Morris. *Mathematical Thought from Ancient to Modern Times*. Oxford University Press, New York. 1972.

Koenigsberger, Leo. *Hermann von Helmholtz*. Translated by Frances A. Welby. Dover Publications, Inc., New York. No date.

Krutch, Joseph Wood. *Human Nature and the Human Condition*. Random House, New York. 1959.

Krutch, Joseph Wood. *The Twelve Seasons*. William Sloane Associates, New York. 1949.

Lang, D. 'Profiles: A Scientist's Advice II' in *New Yorker*. 26 January, 1963.

Layton, Edwin T., Jr. *The Revolt of the Engineers*. The Press of Case Western Reserve University, Cleveland. 1971.

Layton, Edwin T., Jr. 'American Ideologies of Science and Engineering' in *Technology and Culture*. Number 4. October 1976.

Le Corbusier. *Towards a New Architecture*. Translated by Frederick Etchells. The Architectural Press, London. 1927.

Lee, Gerald Stanley. *Crowds*. Doubleday, Page & Company, Garden City. 1913.

Legget, Robert F. *Geology and Engineering*. McGraw-Hill Book Company, Inc., New York. 1962.

Lerner, Eric J. *The Big Bang Never Happened*. Time Books, New York. 1991.

Lerner, Max. *Actions and Passions*. Simon and Schuster, New York. 1949.

Lewis, C.S. *The Pilgrim's Regress: An Allegorical Apology for Christianity, Reason and Romanticism*. Beerdmans, Grand Rapids. 1943.

Lewis, Clarence Irving. *An Analysis of Knowledge and Valuation*. The Open Court Publishing Company, La Salle. 1946.

Lewis, Gilbert N. *The Anatomy of Science*. Yale University Press, New Haven. 1926.

Lewis, Sinclair. *Arrowsmith*. The Modern Library, New York. 1925.

Levy, Hyman. *The Universe of Science*. The Century Company, New York. 1933.

Lichtenberg, George C. *Lichtenberg: Aphorisms & Letters*. Translated by Franz Mautner and Henry Hatfield. Jonathan Cape, London. 1969.

Lieber, Lillian R. *The Education of T.C. MITS*. Norton, New York. 1944.

Lief, Alfred. *The Social and Economic Views of Mr. Justice Brandeis*. The Vanguard Press, New York. 1930.

Lilienthal, David E. *This I Do Believe*. Harper & Brothers Publishers, New York. 1949.

Lilienthal, David E. *TVA: Democracy on the March*. Harper & Brothers Publishers, New York. 1944.

Lippmann, Walter. *A Preface to Politics*. Mitchell Kennerley, New York. 1914.

Lippmann, Walter. *Liberty and the News*. Harcourt, Brace and Howe, New York. 1920.

Locke, John. *An Essay Concerning Human Understanding* in *Great Books of the Western World*. Volume 35. Encyclopædia Britannica, Inc., Chicago. 1952.

Lohr, Lenox, R. *Centennial of Engineering 1852–1952*. Museum of Science and Industry, Chicago. 1953.

Longfellow, Henry Wadsworth. *The Complete Poetical Works of Longfellow*. The Riverside Press, Cambridge. 1920.

Lorenz, Konrad. *On Aggression*. Harcourt, Brace & World, New York. 1966.

Lovins, Amory B. 'Energy Strategy' in *Foreign Affairs*. Volume 55, Number 1. October 1976.

Luminet, Jean-Pierre. *Black Holes*. University Press, Cambridge. 1987.

Mach, Ernst. *The Science of Mechanics*. The Open Court Publishing Co., La Salle. 1942.

Mackay, Charles. *The Collected Songs*. G. Routledge & Co., New York. 1859.

Mailer, Norman. *Of A Fire on the Moon*. Little, Brown and Company, Boston. 1970.

Mangan, James Clarence. *Poems of James Clarence Mangan*. O'Donoghue & Co., Dublin. 1903.

Marlow, Christopher. *Tamburlaine the Great*. Manchester University Press, Manchester. 1981.

Marquardt, Martha. *Paul Ehrlich*. Henry Schuman, New York. 1951.

Marsh, George Perkins. *Man and Nature*. Harvard University Press, Cambridge. 1965.

Marshall, Alfred. *Principles of Economics*. Macmillan and Co., Limited, London. 1898.

Marshall, T.H. 'The Recent History of Professionalism in Relation to Social Structure and Social Policy' in *Canadian Journal of Economic and Political Science*. Volume 5, Number 3, August 1939. University of Toronto Press, Toronto.

Mason, William. *The English Garden: A Poem*. Garland Publishing Co., New York. 1987.

Mathews, J.A. 'Present Tendencies in Engineering Materials' in *Mechanical Engineering*. Volume 48, Number 8. August 1926.

Maxwell, James Clerk. *The Scientific Papers of James Clerk Maxwell*. Volume 2. Edited by W.D. Niven. Dover Publications, New York. 1965.

May, Donald C. 'Choice of Criteria in Applying Mathematical Methods to Evaluation and Analysis of Engineering Systems' in *Journal of Engineering Education*. Volume 40, Number 2. October 1949.

May, Rollo. *The Cry for Myth*. W.W. Norton & Company, New York. 1991.

McArthur, Peter. *The Best of Peter McArthur*. Clarke, Irwin & Company Limited, Toronto. 1967.

McCarthy, Mary. *On the Contrary*. Farrar, Straus and Cudahy, New York. 1961.

McCullough, David. 'Civil Engineers Are People' in *Civil Engineering*. December 1978.

McGonagall, William. *Poetic Gems*. Gerald Duckworth & Co., Ltd., London. 1970.

Meadows, Donella H., Meadows, Dennis L., Randers, Jørgen and Behrens, William W., III. *The Limits to Growth*. Universe Books, New York. 1972.

Mehra, Jagdish (Editor). *The Physicist's Conception of Nature*. Dordrecht, Boston. 1973.

Metsler, William. 'The Cowboy's Lament' in *The Physics Teacher*. February 1977.

Michelmore, Peter. *Einstein: Profile of the Man*. Dodd, Mead, New York. 1962.

Michener, James A. *Space*. Ballentine Books, New York. 1982.

Middendorf, W.H. and Brown, G.T., Jr. 'Orderly Creative Inventing' in *Electrical Engineering*. October 1957.

Milne, A.A. *The House at Pooh Corner*. E.P. Dutton & Co., Inc., New York. 1928.

Milton, John. *Paradise Lost* in *Great Books of the Western World*. Volume 32. Encyclopædia Britannica, Inc., Chicago. 1952.

Moore, A.D. *Invention, Discovery, and Creativity*. Doubleday & Company, Inc., Garden City. 1969.

Morison, George. 'Address at the Annual Convention' in *Transactions of the American Society of Civil Engineers*. June 1895.

Morison, George S. *The New Epoch*. Houghton Mifflin Company, Boston. 1903.

Morley, Christopher. *The Powder of Sympathy*. Doubleday, Page & Company, Garden City. 1923.

Mott-Smith, Morton. *Heat and Its Workings*. D. Appleton and Company, New York. 1933.

Moynihan, Berkeley. *Surgery, Gynecology & Obstetrics*. Volume 31. 1920.

Mumford, Lewis. *Technics and Civilization*. Harcourt, Brace and Company, New York. 1934.

Mumford, Lewis. *The Conduct of Life*. Harcourt, Brace and Company, New York. 1951.

Murphy, Jamie. 'The Quiet Apocalypse' in *Time*. 13 October, 1986.

Murray, Robert F. *Robert F. Murray: His Poems*. Longmans, Green, and Co., London. 1894.

Myrdal, Gunner. *Objectivity in Social Research*. Pantheon Books, New York. 1969.

Newman, James R. *The World of Mathematics*. Volumes III and IV. Simon and Schuster, New York. 1956.

Newton, Sir Isaac. *Mathematical Principles of Natural Philosophy* in *Great Books of the Western World*. Volume 34. Encyclopædia Britannica, Inc., Chicago. 1952.

Nietzsche, Friedrich. *Twilight of the Idols*. Hackett Publishing Company, Inc., Indianapolis. 1997.

Nordenholt, George F. 'A Graduate Can Measure A Bottle' in *Product Engineering*. April 1953.

Nordman, Charles. *Einstein and the Universe*. Henry Holt & Co., New York. 1922.

O'Malley, John R. 'Patents and the Engineer' in *Engineering Facts from Gatorland*. Volume 4, Number 5. December 1967.

O'Neil, W.M. *Fact and Theory*. Sydney University Press, Sydney. 1969.

Oppenheimer, J. Robert. *Science and the Common Understanding*. Simon and Schuster, New York. 1954.

Oppenheimer, Julius Robert. 'The Scientific Foundations for World Order' in *Foundations for World Order*. The University of Denver Press, Denver. 1949.

Oppenheimer, Julius Robert. 'The Tree of Knowledge' in *Harper's Magazine*. Volume 217. October 1958.

Osborn, Alex F. *Applied Imagination*. Charles Scribner's Sons, New York. 1963.

Osgood, W.F. *Mechanics*. The Macmillan Company, New York. 1949.

Page, Ray. *Quote*. Volume 53, Number 20. May 14, 1967.

Panel on Engineering Infrastructure. *Engineering Infrastructure Diagramming and Modeling*. National Academy Press, Washington, D.C. 1986.

Parnes, Sidney J. and Harding, Harold F. *A Sourcebook for Creative Thinking*. Charles Scribner's Sons, New York. 1963.

Parton, H.N. *Science is Human*. University of Otago Press, Dunedin. 1972.

Pascal, Blaise. *Scientific Treatises* in *Great Books of the Western World*. Translated by Richard Scofield. Volume 33. Encyclopædia Britannica, Inc., Chicago. 1952.

Peacock, Thomas Love. *Headlong Hall*. J.M. Dent and Co., London. 1892.

Pearson, Karl. *Life, Letters and Labours of Francis Galton*. Volume II. University Press, Cambridge. No date.

Peirce, Charles S. *Chance, Love, and Logic*. Harcourt, Brace & Company, Inc., New York. 1923.

Penrose, Roger. *The Emperor's New Mind*. Oxford University Press, New York. 1989.

Peter, Lawrence J. 'Peter's People' in *Human Behavior*. August 1976.

Petroski, Henry. *To Engineer is Human*. Cox & Wyman Ltd., Great Britain. 1955.

Philip, Prince Consort of Elizabeth II, Queen of Great Britain. *Men, Machines and Sacred Cows*. Hamish Hamilton, London. 1984.

Piper, Robert J. *Opportunities in an Architecture Career*. Universal Publishing and Distributing Corporation, New York. 1970.

Planck, Max. *Scientific Autobiography and Other Papers*. Philosophical Library, New York. 1949.

Planck, Max. *Where is Science Going?* Translated by James Murphy. W.W. Norton, New York. 1977.

Platonov, Andrei. *Fro and Other Stories*. Progress Publishers, Moscow. 1972.

Poe, Edgar Allan. *The Complete Edgar Allan Poe Tales*. Avenel Books, New York. 1981.

Poincaré, Henri. *The Foundation of Science*. The Science Press, New York. 1921.

Poincaré, Henri. *Science and Hypothesis*. The Walter Scott Publishing Co., Ltd., London. 1905.

Polanyi, Michael. *Personal Knowledge*. The University of Chicago Press, Chicago, Chicago. 1958.

Pólya, G. *How to Solve It*. Princeton University Press, Princeton. 1948.

Pólya, G. *Mathematical Discovery*. John Wiley & Sons, Inc., New York. 1962.

Ponomarev, L.I. *The Quantum Dice*. IOP Publishing, Bristol. 1993.

Pope, Alexander. *The Complete Poetical Works*. Houghton Mifflin and Company, New York. 1903.

Popper, Karl. *Conjectures and Refutations*. Basic Books, Publishers, New York. 1962.

Popper, Karl. *Realism and the Aim of Science*. Rowman and Littlefield, Totowa. 1983.

Popper, Karl. *The Logic of Scientific Discovery*. Hutchinson, London. 1959.

Pottage, John. *Geometrical Investigations*. Addison-Wesley, Reading. 1983.

Queneau, Raymond. *Exercises in Style*. New Direction Publishing Corporation, New York. 1947.

Rae, John A. 'Science and Engineering in the History of Aviation' in *Technology and Culture*. Fall 1961.

Rankine, William John Macquorn. *A Manual of Applied Mechanics*. Charles Griffin and Company, London. 1877.

Read, Opie. *Mark Twain and I*. Reilly & Lee, Chicago. 1940.

Reed, E.G. 'Developing Creative Talent' in *Machine Design*. November 1954.

Reich, Charles A. *The Greening of America*. A Bantam Book, New York. 1970.

Reichen, Charles-Albert. *A History of Astronomy*. Hawthorn Books Inc., New York. 1963.

Renya, Ruth. *The Philosophy of Matter in the Atomic Era*. Asia Publishing House, Bombay. 1962.

Reynolds, Henry T. *Analysis of Nominal Data*. SAGE Publications, Beverly Hills. 1977.

Reynolds, Terry S. 'Defining Professional Boundaries: Chemical Engineering in the Early 20th Century' in *Technology and Culture*. Special Issue. October 1968.

Rickover, H.G. 'A Humanistic Technology' in *Mechanical Engineering*. November 1982.

Roberts, Michael and Thomas, E.R. *Newton and the Origin of Colours*. G. Bell & Sons, London. 1931.

Rogers, G.F.C. *The Nature of Engineering: A Philosophy of Technology*. Macmillan, London. 1983.

Rossman, Joseph. *Industrial Creativity: The Psychology of the Inventor*. University Books, New Hyde Park, New York. 1964.

Roszak, Theodore. *The Making of A Counter Culture*. Doubleday & Company, Inc., Garden City. 1969.

Ruskin, John. *The Stones of Venice*. Volume I. George Routledge & Sons, Limited, London. 1907.

Russell, Bertrand. *Introduction to Mathematical Philosophy*. George Allen and Unwin Ltd., London. 1920.

Russell, Bertrand. *Philosophical Essays*. Longmans, Green, and Co., London. 1910.

Russell, Bertrand. *Sceptical Essays*. W.W. Norton & Company, Inc., New York. 1928.

Russell, Bertrand. *The Impact of Science on Society*. George Allen & Unwin, Ltd., London. 1956.

Russell, Bertrand. *The Scientific Outlook*. George Allen & Unwin Ltd., London. 1931.

Ryle, Gilbert. *The Concept of Mind*. Barnes & Noble, Inc., New York. 1949.

Safonov, V.S. *Courage*. Foreign Languages Publishing House, Moscow. 1953.

Saint Augustine. *De Magistro*. Hackett Publishing Company, Indianapolis. 1995.

Salisbury, J. Kenneth. 'Qualities Industry Wants in Its Engineers' in *General Electric Review*. May 1952.

Samuel, Arthur L. 'Some Moral and Technical Consequences of Automation—A Refutation' in *Science*. Volume 132, Number 3429. September 16, 1960.

Sand, George. *The Haunted Pool*. Translated by Frank Hunter Potter. Shameless Hussy Press, San Lorenzo. 1976.

Saxe, John Godfrey. *The Poetical Works of John Godfrey Saxe*. Houghton, Mifflin and Co., Boston. 1882.

Schlipp, Paul A. *Albert Einstein, Philosopher–Scientist*. Tudor Publishing Co., New York. 1951.

Schumacher, E.F. *Small is Beautiful*. Harper & Row, Publishers, New York. 1973.

Schuyler, Montgomery. 'The Bridge as a Monument' in *Harper's Weekly*. Volume XXVII, Number 1379. 27 May 1883.

Scott, Chas F. 'The Aims of the Society' in *Engineering Education*. Volume 12, Number 3. November 1921.

Seely, Bruce E. 'SHOT, the History of Technology, and Engineering Education' in *Technology and Culture*. Volume 36, Number 4. October 1955.

Seifriz, W. 'A new University' in *Science*. Volume 120, Number 3106. 9 July 1954.

Selye, Hans. *From Dream to Discovery*. McGraw-Hill Book Company, New York. 1964.

Seneca. *The Epistles of Seneca*. Harvard University Press, Cambridge. 1962.

Shakespeare, William. *Hamlet, Prince of Denmark* in *Great Books of the Western World*. Volume 27. Encyclopædia Britannica, Inc., Chicago. 1952.

Shakespeare, William. *Henry IV* in *Great Books of the Western World*. Volume 26. Encyclopædia Britannica, Inc., Chicago. 1952.

Shaw, George Bernard: *Back to Methuselah*. Constable and Company, Ltd., London. 1921.

Shaw, George Bernard. *Man and Superman*. Dodd, Mead & Company, New York. 1945.

Shaw, George Bernard. *The Doctor's Dilemma*. Brentano's Publishers, New York. 1911.

Shipman, T. *Carolina: or, Loyal Poems*. Scholars' Facsimiles & Reprints, Delmar. 1980.

Shute, John. *The First and Chief Groundes of Architecture*. Reprinted by The Gregg Press Limited, London. 1960.

Shute, Nevil. *Slide Rule*. William Heinemann, Ltd., Melbourne. 1954.

Shuyler, Montgomery. 'The Bridge as a Monument' in *Harper's Weekly*. 27 May 1883.

Siegel, Eli. *Damned Welcome*. Definition Press, New York. 1972.

Sigerist, Henry E. *A History of Medicine*. Volume I. Oxford University Press, New York. 1951.

Simon, Herbert. *Models of Man: Social and Rational*. John Wiley & Sons, Inc., New York. 1957.

Simon, Herbert A. *Administrative Behavior*. The Macmillan Company, New York. 1947.

Simpson, N.F. *New English Dramatists 2*. Penguin Books, Ltd., Harmondsworth. 1960.

Singer, Charles. *Studies in the History and Method of Science*. William Dawson & Sons, Ltd., London. 1955.

Smith, George Otis. 'What are Facts?' in *Civil Engineer*. Volume 2, Number 3. March 1932.

Smith, R.B. 'Professional Responsibility of Engineering' in *Mechanical Engineering*. Volume 86, Number 1. January 1964.

Smith, R.B. 'Engineering is . . .' in *Mechanical Engineering*. Volume 86, Number 5. May 1964.

Smith, Ralph J. *Engineering as a Career*. McGraw-Hill Book Company, Inc., New York. 1962.

Snow, C.P. *The Two Cultures: and A Second Look*. At the University Press, Cambridge. 1964.

Solzhenitsyn, Aleksandr. *The First Circle*. Harper and Row, Publishers, New York. 1968.

Sophocles. *Antigone* in *Great Books of the Western World*. Translated by Sir Richard C. Jebb. Volume 5. Encyclopædia Britannica, Inc., Chicago. 1952.

Spencer, Herbert. *The Principles of Biology*. Volume I. D. Appleton and Company, New York. 1886.

Spencer-Brown, George. *Laws of Form*. Allen & Unwin, London. 1969.

Spengler, Oswald. *The Decline of the West*. Alfred A. Knopf, New York. 1934.

Spiegel, M.R. 'E and M' in *Mathematics Magazine*. Volume 54, Number 3. May 1981.

Sporn, Philip. *Foundations of Engineering*. The Macmillan Company, New York. 1964.

Steinman, D.B. 'SUSPENSION BRIDGE' in *American Engineer*. February 22–28, 1953.

Steinmetz, Charles Proteus. *Engineering Mathematics*. McGraw-Hill Book Company, New York. 1911.

Stenger, Victor J. *Not By Design*. Prometheus Books, Buffalo. 1988.

Stern, Lawrence. *Tristram Shandy*. The Modern Library, New York. 1928.

Stewart, Ian. *Does God Play Dice?* Basil Blackwell Inc., Cambridge. 1990.

Stevenson, Adlai E. 'My Faith in Democratic Capitalism' in *Fortune Magazine*. October 1955.

Stevenson, Robert Louis. 'Records of a Family of Engineers' in *Letters and Miscellanies of Robert Louis Stevenson*. Charles Scribner's Sons, New York. 1898.

Swift, Jonathan. *Gulliver's Travels* in *Great Books of the Western World*. Volume 36. Encyclopædia Britannica, Inc., Chicago. 1952.

Swift, Jonathan. *Satires and Personal Writings*. Oxford University Press, New York. 1932.

Szent-Györgyi, Albert. 'Teaching and the Expanding Knowledge' in *Science*. Volume 146, Number 3649. December 1964.

Tagore, Rabindranath. *Collected Poems and Plays of Rabindranath Tagore*. Macmillan, New York. 1937.

Taylor, Bert. *The So Called Human Race*. A.A. Knopf, New York. 1922.

Taylor, E.S. 'Report on Engineering Design' in *Journal of Engineering Education*. Volume 51, Number 8. April 1961.

Tennyson, Alfred. *The Complete Poetical Works of Tennyson*. Houghton Mifflin Company, Boston. 1898.

Terence. *Phormio*. Translated by Frank O. Copley. The Liberal Arts Press, New York. 1958.

Thompson, D'Arcy. *On Growth and Form*. Volume I. Cambridge University Press, London. 1959.

Thomson, Sir George. *The Inspiration of Science*. Oxford University Press, London. 1961.

Thomson, William [Baron Kelvin]. *Baltimore Lectures on Molecular Dynamics and the Wave Theory of Light*. C.J. Clay and Sons, London. 1904.

Thomson, William [Baron Kelvin]. *Popular Lectures and Addresses*. Macmillan and Co., London. 1891–94.

Thoreau, Henry David. *Walden*. Published for the Classics Club by W.J. Black, New York. 1942.

Thoreau, Henry David. *The Cambridge History of American Literature in Four Volumes*. G.P. Putnam's Sons, New York. 1918.

Thring, M.W. 'On the Threshold' in *Proceedings of the Institution of Mechanical Engineers*. Volume 179, Part I. 1963–64.

Toffler, Alvin. *Future Shock*. Random House, New York. 1970.

Trefethen, Joseph M. 'Geology for Civil Engineers' in *Journal of Engineering Education*. Volume 39, Number 7. March 1949.

Truesdell, Clifford A. *Six Lectures on Modern Natural Philosophy*. Springer-Verlag, Berlin. 1966.

Tukey, John W. 'The Future of Data Analysis' in *Annals of Mathematical Statistics*. Volume 33, Number 1. March 1962.

Twain, Mark. *Following the Equator*. The American Publishing Company, Hartford. 1847.

Twain, Mark. *The American Claimant*. Harper and Brothers, New York. 1924.

Updike, John. *Telephone Poles and Other Poems*. Alfred A. Knopf, New York. 1964.

Vallery-Radot, René. *The Life of Pasteur*. Volume II. Translated by Mrs. R.L. Devonshire. McClure, Phillips and Co., New York. 1902.

van Noordwijk, J. 'The Bioassayist' in *Perspectives in Biology and Medicine*. Volume 29, Number 2, Winter 1986.

Vickers, Brian. *Francis Bacon*. Oxford University Press, Oxford. 1996.

Vincenti, Walter G. *What Engineers Know and How They Know It*. The Johns Hopkins University Press, Baltimore. 1990.

Vitruvius. *Vitruvius on Architecture*. Volume I. Translated by Frank Granger. William Heinemann Ltd., London. 1931.

Vollmer, James. 'Engineering. Growing, Steady State, or Evanescent' in *The Bridge of Eta Kappa Nu*. Volume 65, Number 4. August 1969.

von Karman, Theodore. 'Creativity is a Task, Not a Trait' in *Machine Design*. May 25, 1967.

Waddell, J.A.L., Skinner, Frank W. and Wessman, Harold E. *Vocational Guidance in Engineering Lines*. The Mack Printing Company, Easton. 1933.

Walker, Eric. 'Engineers and/or Scientists' in *Journal of Engineering Education*. Volume 51. February 1961.

Walker, Eric A. 'Our Tradition-Bound Colleges' in *Engineering Education*. October 1969.

Walker, William H. *High Lights of Fifty Years*. American Institute of Chemical Engineers, New York. 1958.

Wallace, Lew. *The Prince of India*. Grosset & Dunlap Publishers, New York. 1893.

Weaver, Jefferson Hane. *The World of Physics*. Volume II. Simon and Schuster, New York. 1987.

Weinberg, Gerald M. *Rethinking Systems Analysis and Design*. Little, Brown & Co., Inc., Boston. 1982.

Wellington, Arthur Mellen. *The Economic Theory of the Location of Railways*. John Wiley & Sons, New York. 1900.

Wells, H.G. *Mr. Britling Sees It Through*. Cassell and Company, Limited, London. 1916.

Western, David and Pearl, Mary C. *Conservation for the Twenty-first Century*. Oxford University Press, New York. 1989.

Weyl, Hermann. *Symmetry*. Princeton University Press, Princeton. 1952.

Weyl, Hermann. *The Theory of Groups and Quantum Mechanics*. Methuen & Co., Ltd., London. 1931.

Whipple, E.P. *Literature and Life*. Houghton, Mifflin and Company, Boston. 1871.

Whitehead, Alfred North. *An Introduction to Mathematics*. Oxford University Press, London. 1948.

Whitehead, Alfred North. *Process and Reality*. The Macmillan Company, New York. 1929.

Whitehead, Alfred North. *Science and the Modern World*. The Free Press, New York. 1967.

Whitehead, Alfred North. *The Aims of Education*. The Macmillan Company, New York. 1929.

Whitehead, Alfred North. *The Organization of Thought*. Williams and Norgate, London. 1917.

Whittier, John. *The Complete Poetical Works of John Greenleaf Whittier*. Houghton Mifflin Company, Boston. 1927.

Wigner, Eugene P. *Symmetries and Reflections*. Indiana University Press, Bloomington. 1967.

Wilczek, Frank and Devine, Betsy. *Longing for the Harmonies*. W.W. Norton & Company, New York. 1988.

Wilde, Oscar. *Phrases and Philosophies for the Use of the Young*. A.R. Keller & Co., London. 1907.

Wilde, Oscar. *The Importance of Being Earnest*. Appeal to Reason, Girard. 1921.

Wilde, Oscar. *The Picture of Dorian Gray*. The World Publishing Co., Cleveland. 1946.

Wilkins, John. *The Discovery of a World in the Moone*. London. 1638.

Wilson, John. *The Cheats*. Basil Blackwell, Oxford. 1935.

Winsor, Dorothy A. *Writing Like an Engineer*. Lawrence Erlbaum Associates, Publishers, Mahwah. 1996.

Wittgenstein, Ludwig. *Remarks on the Foundations of Mathematics*. Translated by G.E.M. Anscombe. B. Blackwell, Oxford. 1956.

Wittgenstein, Ludwig. *Tractatus Logico-Philosophicus*. Routledge, London.

Wolfle, Dael. *Symposium on Basic Research*. American Association for the Advancement of Science, Washington, D.C. 1959.

Woodson, Thomas T. *Introduction to Engineering Design*. McGraw-Hill Book Company, New York. 1966.

Woolf, Harry. *Some Strangeness in the Proportion*. Addison-Wesley Publishing Company, Reading. 1980.

Wooten, Henry. *The Elements of Architecture*. Da Capo Press, Amsterdam. 1970.

Wright, Frank Lloyd. *Frank Lloyd Wright: An Autobiography*. Duell, Sloan and Pearce, New York. 1943.

Wright, Frank Lloyd. 'Frank Lloyd Wright and His Art' in *New York Magazine*. October 4, 1953.

Wright, Harold Bell. *The Winning of Barbara Worth*. The Book Supply Company, Chicago. 1911.

Zener, C. 'Engineering in the Future' in *Florida Engineer*. October 1965.

PERMISSIONS

Grateful acknowledgement is made to the following for their kind permission to reprint copyright material. Every effort has been made to trace copyright ownership but if, inadvertently, any mistake or omission has occurred, full apologies are herewith tendered.

Full references to authors and the titles of their works are given under the appropriate quotation.

A HISTORY OF MEDICINE by Henry E. Sigerist. Volume I. Copyright 1951. Reprinted by permission of Oxford University Press, Oxford, UK.

AN INTRODUCTION TO MATHEMATICS by Alfred North Whitehead. Copyright 1948. Reprinted by permission of Oxford University Press, Oxford, UK.

BLAMING TECHNOLOGY by Samuel C. Florman. Copyright 1981. Reprinted by permission of St Martin's Press, Inc., New York.

BRAIN OF THE FIRM by Stafford Beer. Copyright 1981. Reprinted by permission of the publisher John Wiley & Sons, Ltd., Chichester, UK.

CONJECTURES AND REFUTATIONS by Karl Popper. Reprinted by permission of Melitta Mew and Alfred Raymond Mew, Surrey, UK.

CONSERVATION FOR THE TWENTY-FIRST CENTURY by David Western and Mary C. Pearl. Copyright 1989. Reprinted by permission of Oxford University Press, Oxford, UK.

CONVERSATIONS WITH JORGE LUIS BORGE edited by Richard Durgin. Copyright 1968 by Henry Holt and Company, Inc. Reprinted by permission of the publisher Henry Holt and Company, Inc.

COSMIC CANTICLE by Ernesto Cardenal. Copyright 1993. Reprinted by permission of the publisher, Curbstone Press, Willimantic, Connecticut.

CREATIVE SYNTHESIS IN DESIGN by John R.M. Alger and Carl V. Hays. Copyright 1964. Reprinted by permission of the publisher Prentice-Hall, Inc., Englewood Cliffs, New Jersey.

SILENCE 1961 by John Cage. Copyright 1961. Reprinted by permission of the publisher University Press of New England, Hanover, New Hampshire.

STATISTICS FOR ECONOMISTS by Roy George Douglas Allen. Copyright 1949. Reprinted by permission of the publisher, Routledge, Andover, UK.

'American Ideologies Of Science And Engineering' in TECHNOLOGY AND CULTURE by Edwin T. Layton Jr. Copyright 1976. Reprinted by permission of the publisher The University of Chicago Press, Chicago, Illinois.

'Engineering in the 20th Century' in TECHNOLOGY AND CULTURE by Robert Friedel. Copyright 1968. Reprinted by permission of the publisher The University of Chicago Press, Chicago, Illinois.

'Science and Engineering in the History of Aviation' in TECHNOLOGY AND CULTURE by John A. Rae. Copyright 1961. Reprinted by permission of the publisher The University of Chicago Press, Chicago, Illinois.

TEN NIGHTS IN A BAR-ROOM AND WHAT I SAW THERE by T.S. Arthur. Edited by Donald A. Koch. Copyright 1964 by the President and Fellows of Harvard College. Reprinted by permission of Harvard University Press.

THE ASCENT OF MAN by Jacob Bronowski. Copyright 1973. Reprinted by permission of the publisher Little, Brown and Company, New York.

THE CIVILIZED ENGINEER by Samuel C. Florman. Copyright 1987. Reprinted by permission of St Martin's Press, Inc., New York.

THE COMMON SENSE OF SCIENCE by J. Bronowski. Copyright 1953. Reprinted by permission of Harvard University Press.

THE COMPLETE POETICAL WORKS OF WILLIAM COWPER by William Cowper. Copyright 1913. Reprinted by permission of Oxford University Press, Oxford, UK.

THE COUNTRY PARSON, THE TEMPLE by George Herbert. Copyright 1981. Reprinted by permission of the Paulist Press, New York.

THE EDUCATION OF HENRY ADAMS by Henry Adams. Copyright 1946. Reprinted by permission of the publisher Random House, New York.

THE EMERGENCE OF PROBABILITY by Ian Hacking. Copyright 1975. Reprinted by permission of the publisher Cambridge University Press, Cambridge, UK.

SUBJECT BY AUTHOR INDEX

-A-

analysis

Allen, Roy George Douglas
Not even the most subtle and skillful analysis can overcome completely the unreliability of basic data..., 1

Amiel, Henri Frédéric
Analysis kills spontaneity..., 1

Minds accustomed to analysis..., 1

Keensy, Ralph
...be wary of analysts that try to quantify the unquantifiable..., 1

Poe, Edgar Allan
...the ingenious man is often remarkably incapable of analysis..., 1

Reed, E.G.
...the primary purpose of analysis..., 2

Whitehead, Alfred North
It requires a very unusual mind to undertake the analysis of the obvious..., 2

Woodson, Thomas T.
The engineer uses all the analysis...he can command..., 92

answer

Arnold, John E.
...there is no one right answer to creative problems..., 3

Berkeley, Edmund C.
The moment you have worked out an answer..., 3

Dobie, J. Frank
Putting on the spectacles of science..., 3

Hodnett, Edward
You have to ask a precise question to get a precise answer..., 3

Woodson, Thomas T.
...an unproven answer is wrong..., 3

answers

Adams, Henry Brooks
Unintelligible answers to insoluble problems..., 3

Baez, Joan
...hypothetical questions get hypothetical answers..., 3

arch

Fergusson, James
...an arch never sleeps..., 4

architect

Alexander, Daniel Asher
...I am not a builder, I am an architect..., 5

geology
Dawkins, Boyd
  Geology stands to engineering...,
    140
Trefethen, Joseph M.
  ...relationship between civil
    engineering and geology...,
    140
goal
Farber, Eric A.
  ...striving toward a
    definite...goal..., 141
goals
Fredrickson, A.G.
  Only by setting up defined
    goals..., 141
  We must try to set up some
    definite goals..., 141
graphic
Anthony, Gardner C.
  ...one of the many graphic
    representations of those
    ideas..., 142
graphics
Rogers, Will
  ...We learn through the eye and
    not the noggin..., 142
gravity
Bronowski, Jacob
  ...gravity is vertical..., 143
Emerson, Ralph Waldo
  ...learning the action of light,
    motion, gravity..., 143
Unknown
  What goes up must come
    down..., 143

-H-
heat
Flanders, Michael
  You can't pass heat from a
    cooler to a hotter..., 144
Keane, Bill

Heat makes things expand...,
  144
Mayer, Julius Robert von
  Concerning the intimate nature
    of heat..., 144
McNeil, I.
  Let there also be heat..., 144
Metsler, William
  Heat's what it's supposed to
    make..., 145
Mott-Smith, Morton
  People are always exaggerating
    temperatures..., 145
Poincaré, Henri
  ...the word heat devoted many
    generations to error..., 145

-I-
idea
Bagehot, Walter
  ...the pain of a new idea..., 146
Bernard, Claude
  If an idea presents itself to us...,
    146
Butler, Samuel
  Every idea has something
    of the pain and peril of
    childbirth..., 146
Eliot, George
  ...the moment of finding an
    idea..., 147
Eliot, T.S.
  Between the idea and the
    reality..., 147
James, William
  An idea, to be suggestive..., 148
Locke, John
  ...to invent or frame one new
    simple idea..., 148
Milne, A.A.
  ...a Thing which seemed very
    Thingish inside you..., 148
Penrose, Roger
  A beautiful idea..., 17

If I can make a mechanical
    model, I understand it..., 195
models
Karlin, Samuel
  The purposes of models..., 195
Unknown
  ...they mainly make models...,
    195
motion
Cohen, I. Bernard
  ...most people's views about
    motion..., 196
Galilei, Galileo
  There is, in nature, perhaps
    nothing older than motion...,
    196
Jeans, Sir James Hopwood
  The motions of the electrons...,
    196
motions
Wittgenstein, Ludwig
  ...we can describe the motions of
    the world using Newtonian
    mechanics..., 197
motto
US Army Corps of Engineers
  Let's try, 90

-O-

observation
Aurelius, Marcus [Antoninus]
  Consider that everything which
    happens, happens justly,
    and if thou observest
    carefully, thou wilt find it to
    be so, 198
Ayres, C.E.
  ...Moses...announced that his
    laws were based on direct
    observation..., 198
Bachelard, Gaston
  A scientific observation is
    always a committed
    observation..., 198

Bernard, Claude
  Observation, then, is what
    shows facts..., 199
Blake, William
  A fool sees not the same tree
    that a wise man sees, 199
Bohr, Niels
  ...the foundation on which the
    customary interpretation of
    observation was based...,
    199
Carlyle, Thomas
  ...the more surprising that we do
    not look round a little..., 199
Eddington, Sir Arthur
  ...observation is the supreme
    Court of Appeals..., 200
Emerson, Ralph Waldo
  ...there is a great difference in
    the beholders..., 200
Greer, Scott
  ...the link between observation
    and formulation..., 200
Hooke, Robert
  ...the soundness of
    Observations..., 201
Hutten, Ernest H.
  ...certain conditions under
    which the observable thing
    is perceived are tacitly
    assumed..., 201
O'Neil, W.M.
  ...the solid ground of
    observation..., 201
Poincaré, Henri
  ...we must use our
    observations..., 202
Saxe, John Godfrey
  ...that each by observation might
    satisfy his mind..., 202
Unknown
  I am an observation..., 202
Whitehead, Alfred North
  'tis here, 'tis there, 'tis gone...,
    203

-T-

table
Playfair, William
...a man who has carefully investigated a printed table..., 245

tables
Benzecri, J.P.
...these long lists of little tables..., 245
Carlyle, Thomas
Tables are like cobwebs..., 245

technical
Orwell, George
Men are only so good as their technical developments allows them to be, 246

technical skills
Hoffer, Eric
Where there is the necessary technical skill to move mountains..., 246

technics
Mumford, Lewis
...however far modern science and technics have fallen short..., 253

technological
Hanham, H.J.
Great technological advances are always around the corner, 246

technological advance
Harrington, Michael
If there is technological advance..., 246

technological innovation
Cardwell, D.S.L.
There are...a number of different modes of technological innovation..., 249
Fredrickson, A.G.
...if a technological innovation has a good side..., 155

technological invention
DeSimone, Daniel V.
Technological invention and innovation are the business of engineering..., 249

technological surge
Soule, Michael
...swept along by this vast technological surge..., 255

technological world
Oppenheimer, Julius Robert
...ever more specialized and expert technological world..., 253

technology
Adler, Alfred
The confusion of science with technology is understandable..., 247
Allen, Charles M.
If the human race wants to go to hell...technology can help it get there..., 247
Ashby, Sir Eric
The habit of apprehending a technology in its completeness..., 247
Association of American Colleges
...unable to comprehend the the technology we invent..., 248
Ballard, J.G.
Science and technology multiply around us..., 248
Barzun, Jacques
When techne assails the senses..., 248
Beer, Stafford
If it works, it's out of date..., 248
Bronowski, Jacob
Every civilization has been grounded on technology..., 248
Bunge, M.
...technology knows in order to elicit changes, 248

AUTHOR BY SUBJECT INDEX

-C-

cause and effect, 28
knowledge, 170
logic, 177
Holmes, Sherlock [*creation of
    Arthur Conan Doyle
    (1859–1930) British novelist*]
decision, 44
ideas, 147
method, 193
observe, 201
problem, 216
Holton, G.
discovery, 49
Hood, Thomas (1799–1845)
English poet
engineer, 70
Hooke, Robert (1635–1703)
English physicist
measurement, 188
observation, 201
Hoover, Herbert (1874–1964)
US president
engineer, 70
engineering, 100
specifications, 238
Hoover, T.J.
engineering, 101
Howland, W.E.
engineers, 71
Hoyle, Fred (1915–)
British astronomer
creative, 39
Hubbard, Elbert (1856–1915)
US author/editor
inventors, 159
logic, 177
Hugo, Victor (1802–1885)
French poet
beautiful, 17
blunders, 113
inspiration, 156
Hume, David (1711–1776)
Scottish philosopher
experience, 120
Hutten, Ernest H. (1908–)

observation, 201
Huxley, Aldous (1894–1963)
English novelist
engineers, 101
experience, 120
facts, 128
inventions, 159
knowledge, 170
technological progress, 251
theorizing, 257
Huxley, Julian (1887–1975)
English biologist
fact, 128
Huxley, Thomas H. (1825–1895)
English naturalist
decision, 44
engineer, 72
errors, 113
fact, 129
failure, 131
ideas, 147
knowledge, 171
laws, 173
prayer, 210

-I-
Isaiah 40:12, 42:20
measured, 188
observe, 201

-J-
Jackson, Hughlings
cause and effect, 28
James, William (1842–1910)
American philosopher
data, 43
idea, 148
Jeans, Sir James Hopwood
    (1877–1946)
English astronomer
engineer, 101
motion, 196

Rossman, Joseph
  idea, 148
  inventors, 165
Roszak, Theodore
  problem, 218
Ruskin, John (1819–1900)
English writer
  architect, 8
  architecture, 12
  build, 21
  builder, 24
  buildings, 25
  failure, 131
Russell, Bertrand (1872–1970)
English philosopher/mathematician
  common sense, 31
  logic, 178
  machines, 181
  prayer, 211
  truth, 265
Rutherford, Ernest (1871–1937)
British physicist
  discovery, 50
Ryle, Gilbert (1900–1976)
British philosopher
  machine, 182

-S-
Safonov, V.
  discovery, 50
Sagan, Carl (1934–1996)
US physicist
  engineer, 78
Saint Augustine (354–430)
Latin religious writer
  formula, 136
  time, 260
Salisbury, J. Kenneth
  engineers, 78
  perspective, 207
Samuel, Arthur L.
  machine, 182
Sand, George (1804–1876)
French novelist

  engineer, 77
Sartre, Jean-Paul (1905–80)
French novelist/philosopher
  beauty, 17
Saxe, John Godfrey (1816–1887)
  observation, 202
Schelling, Friedrich (1775–1854)
German philosopher
  architecture, 12
Schiller, F.C.S. (1864–1937)
  logic, 178
Schlicter, Dean
  mathematics, 185
Schumacher, E.F. (1911–1977)
  machines, 182
  technology, 254
Schuyler, Montgomery
  bridge, 19
Scott, Chas F.
  engineering, 105
Seeger, Peggy (1935)
US singer
  engineer, 78
Seely, Bruce E. (1953–)
  engineer, 78
Seifriz, W.
  communication, 33
Seneca (8 BC–65 AD)
Roman philosopher
  builders, 24
Shakespeare, William (1564–1616)
English poet
  build, 21
  engineer, 78
  reasons, 223
Shaw, George Bernard (1856–1950)
Irish dramatist
  create, 37
  engineer, 78
  problems, 219
Shewhart, W.A. (1891–1967)
  engineering, 105
Shipman, T. (1632–1680)
  architect, 8